T0299062

AN OUTLINE OF ERGODIC THEORY

This informal introduction provides a fresh perspective on isomorphism theory, which is the branch of ergodic theory that explores the conditions under which two measure-preserving systems are essentially equivalent. It contains a primer in basic measure theory, proofs of fundamental ergodic theorems, and material on entropy, martingales, Bernoulli processes, and various varieties of mixing.

Original proofs of classic theorems – including the Shannon–McMillan–Breiman theorem, the Krieger finite generator theorem, and the Ornstein isomorphism theorem – are presented by degrees, together with helpful hints that encourage the reader to develop the proofs on their own. Hundreds of exercises and open problems are also included, making this an ideal text for graduate courses. Professionals needing a quick review, or seeking a different perspective on the subject, will also value this book.

STEVEN KALIKOW is a Visiting Professor in the Department of Mathematical Sciences at the University of Memphis.

RANDALL MCCUTCHEON is Associate Professor in the Department of Mathematical Sciences at the University of Memphis.

CAMBRIDGE STUDIES IN ADVANCED MATHEMATICS

Editorial Board:

B. Bollobás, W. Fulton, A. Katok, F. Kirwan, P. Sarnak, B. Simon, B. Totaro

All the titles listed below can be obtained from good booksellers or from Cambridge University Press. For a complete series listing visit: http://www.cambridge.org/series/sSeries.asp?code=CSAM

Already published

An Outline of Ergodic Theory

STEVEN KALIKOW
RANDALL MCCUTCHEON

University of Memphis

CAMBRIDGE
UNIVERSITY PRESS

Shaftesbury Road, Cambridge CB2 8EA, United Kingdom

One Liberty Plaza, 20th Floor, New York, NY 10006, USA

477 Williamstown Road, Port Melbourne, VIC 3207, Australia

314–321, 3rd Floor, Plot 3, Splendor Forum, Jasola District Centre, New Delhi – 110025, India

103 Penang Road, #05–06/07, Visioncrest Commercial, Singapore 238467

Cambridge University Press is part of Cambridge University Press & Assessment, a department of the University of Cambridge.

We share the University's mission to contribute to society through the pursuit of education, learning and research at the highest international levels of excellence.

www.cambridge.org
Information on this title: www.cambridge.org/9780521194402

First published 2010

A catalogue record for this publication is available from the British Library

Library of Congress Cataloging-in-Publication data
Kalikow, Steven, 1950–
An outline of ergodic theory / Steven Kalikow, Randall McCutcheon.
p.   cm. – (Cambridge studies in advanced mathematics ; 122)
ISBN 978-0-521-19440-2 (hardback)
1. Ergodic theory.   2. Isomorphisms (Mathematics)   I. McCutcheon,
Randall, 1965–   II. Title.   III. Series.
QA313.K35   2010
515′.48–dc22
2010000325

ISBN   978-0-521-19440-2   Hardback

Contents

Preface

This book treats *isomorphism theory* – that branch of ergodic theory dealing with the question of when two measure-preserving systems are, in a certain sense, essentially equivalent. Although these topics have received fair treatment in several books, we think that the time is right for a fresh perspective. Indeed, with ergodic theory becoming more fashionable in its connections with number theory and additive combinatorics, yet also more abstract and structure-laden, it is interesting to observe the extent to which progress in its original concerns, classification of measure-preserving systems up to isomorphism, was achieved via combinatorial/probabilistic reasoning. Our hope is that the ergodic theory revival currently underway will find its way to isomorphism theory, and revitalize it as well.

We have also attempted to write a book that teaches general mathematical thinking in a unique manner. Most graduate level textbooks in pure mathematics provide detailed proofs of theorems followed by exercises. We[1] have tried to write this book in such a way as to make the proofs of the theorems themselves the exercises. Optional details, of which readers may want more or less, may be relegated to footnotes or to sections labeled "Remark" or "Comment".

Indeed, proofs of major theorems are generally presented twice; once labeled "Idea of proof", in which the reader is called on to flesh out the argument from a very basic outline, then again with the label "Sketch of proof", in which more details are given. We consider it important that the reader attempt to work through the "Idea" section before or instead of the "Sketch". This reading strategy benefits:

(a) the serious beginner wishing to work out the proofs on their own, with helpful guidance from the book;
(b) the student wanting a basic overview of ergodic theory, who doesn't want to be overburdened with notation while trying to understand the ideas;
(c) the professional mathematician who is familiar with the material, but wants a quick review or different perspective on it.

---

[1] The first author has circulated a rough precursor to this book as an MS Word document on the Internet for a number of years; the pedagogic philosophy employed here (and all original proofs of major theorems) are due to him. R.M.

This book differs from other books of its type not only in the careful compartmentalization of detail; much in our approach to classical theorems (those of Shannon–McMillan–Breiman and Ornstein, for example) would be considered unorthodox, however presented. Indeed, our aim is not to present "distinctively modern", "slick" or "book" proofs. Our aim is to present *our* proofs, stumbled upon by active engagement with the subject matter. Our hope is that readers will follow our example, and come to favor their own proofs as well.

Introduction

Note: this introduction is written in an intuitive style, so a scientifically oriented non-mathematician might get something out of it. It is the only part of the book that requires no mathematical expertise.

Question: *What is ergodic theory?*

Let's start with two examples.

Example 1: Imagine a potentially oddly shaped billiard table having no pockets and a frictionless surface. Part of the table is painted white and part of the table is painted black. A billiard ball is placed in a random spot and shot along a trajectory with a random velocity. You meanwhile are blindfolded and don't know the shape of the table. However, as the billiard ball careens around, you receive constant updates on when it's in the black part of the table, and when it's in the white part of the table. From this information you are to deduce as much as you can about the entire setup: for example, whether or not it is possible that the table is in the shape of a rectangle.

Example 2: (This example is extremely vague by intention.) Imagine you are receiving a sequence of signals from outer space. The signal seems to be in some sense random, but there are recurring patterns whose frequencies are stationary (that is, do not alter over time). We are unable to detect a precise signal but we can encode it by interpreting five successive signals as one signal: unfortunately, this code loses information. Furthermore, we make occasional mistakes. We wish to get as much knowledge as possible about the original process.

Measure preserving transformations. The subject matter encompassing the previous two examples is called ergodic theory. Ergodic theory models situations (like the examples) under the abstraction of *measure-preserving transformations*. To understand that concept, we need to understand what *measures* are and what *transformations* are.

A measure is a concept of size that tells you how big a set is, or, in the language of probability, how probable an event is. (A probability space is a space whose total measure is equal to 1.) It has to act like a notion of size should: the measure of the union of two disjoint sets has to equal the sum of the measures of the sets, for example. A transformation is a way of mapping a space to itself by assigning one point to another. In many modeling applications, the transformation indicates evolution in time: for example, it may map the position and direction of a billiard ball at the current time to the position and direction of the ball one second later.

Ergodic theory is the study of transformations on probability spaces that preserve measure. So, for example, if a set A of points has measure $\frac{1}{3}$, and a transformation T is measure-preserving, then the set of points which are mapped into A by T will also have measure $\frac{1}{3}$. When the measure is interpreted as a probability, the measure-preserving property indicates the stationarity or time-invariance of the expected frequencies of certain events (like the probability that the billiard ball of Example 1 lies in the white part of the table).

Processes. When you apply a transformation over and over, checking after each application whether some event has occurred and recording the result, you get a *process*. The language of processes and the language of transformations are really just two different ways of describing the same thing. You can get a process out of Example 1 if you record at one-second intervals, by writing down either B or W, the location of the moving billiard ball. For example, the output $BBWWB\ldots$ represents the ball having been in the black, black, white, white and black areas at times 0, 1, 2, 3 and 4, respectively. You can also get processes out of real world systems, without foreknowledge of any transformation acting. For example, say you record each day at noon whether it's rainy or sunny by writing down R or S. If you did this every day into both the future and the past, you would output a doubly infinite string of Rs and Ss, thus: $\ldots SRRS(S)RRRS\ldots$ Here the parentheses identify the current day (it is sunny today, will be rainy tomorrow, and was sunny yesterday, etc.).

To transfer this example to the language of transformations, note that the set of all doubly infinite strings of Rs and Ss forms a space, and a natural transformation of this space is the *shift*, which moves time forward one day. (Hence the shift takes the above sequence to $\ldots SRRSS(R)RRS\ldots$) An appropriate measure can be derived from the probabilities of rain and sun respectively on the various days. This measure will be preserved by the shift precisely when the original process was stationary.

In this book, we will usually study measure-preserving transformations using the language of stationary processes. Here is a summary of the important concepts and theorems we will cover.

(1) *Isomorphism*: Suppose that in Example 1 we were to change which part of the table is painted white and which is painted black. Then you would have a different process. But our new process could end up being equivalent to the original process in the sense that if you know the output of either process infinitely far into both the past and future, it would tell you the

output of the other process. Very roughly, one says that two processes are *isomorphic* when there is a nice way to map outputs of one to outputs of the other so that they determine each other. In general, determining whether there is such a map can be nearly impossible; much of this book is about ways to do it in a few cases.

(2) *Ergodicity*: An *ergodic* process or transformation is one that cannot be expressed as a combination of two simpler processes (or transformations). For example, consider the process that picks a random person and then spits out an enormous sequence of Ls and Rs according to which hand that person uses to twist open all the doorknobs they encounter their whole life. That process is certainly not going to be ergodic because the character of the output will be divided in a predictable way according to whether the person chosen is left-handed or right-handed. Assuming the proportion of left-handed people in the general population to be 0.09, the whole process would then be expressible as 0.09 (left-hand process) $+ 0.91$ (right-hand process).

(3) *Birkhoff ergodic theorem*: When an ergodic transformation is repeatedly applied to form an ergodic process, then with probability 1, the frequency of time an output of that process spends in a given set is the measure of that set, e.g. if the measure of a set is $\frac{1}{3}$, then it will spend (in an asymptotic sense) $\frac{1}{3}$ of the time in that set.

(4) *Rohlin tower theorem*: Fix an arbitrary positive integer, say 678. For any measure-preserving transformation T that does not simply rotate finite sets of points around, you can break almost the whole space into 678 equally sized disjoint sets $A_1, \ldots, A_{678}$ such that if you arrange the sets as the rungs of a ladder, the transformation consists in simply walking up the ladder; that is, $T A_i = A_{i+1}$.

(5) *Shannon–McMillan–Breiman theorem*: Consider an ergodic process that spits out doubly infinite strings of as and bs. If you pick a random doubly infinite string, then with probability 1, when you look at its sequence of finite initial strings (e.g. $a, ab, abb, abba, \ldots$), that sequence will have probabilities that asymptotically approach a fixed rate of exponential decay. Moreover, that rate of decay will not depend on the sequence you choose.

(6) *Entropy*: The exponential rate of decay just mentioned is called the *entropy* of the process. Recall that essentially all of the doubly infinite strings have the same exponential decay rate. Call the ones that do *reasonable names*. Then the number of reasonable names is approximately equal to the reciprocal of this exponentially decaying probability, that is, it is a quantity that increases at a fixed exponential rate. Thus entropy can also be thought of as

being the asymptotic exponential growth rate of the number of reasonable names.

(7) *Kolmogorov entropy invariance*: Any two isomorphic processes must have the same entropy. This provides a quick way to identify two processes as *not* being isomorphic, namely, having different entropies.

(8) *Independent process*: A stationary process on an alphabet in which the next letter to come in the output string is always independent of the ones that came previously is called an *independent process*. For example, repeatedly rolling a die (even a loaded die) gives an independent process.

(9) *Ornstein isomorphism theorem*: Says that two stationary independent processes are isomorphic if and only if they have the same entropy. Indeed, the standard proofs of the theorem say even more, as they give a condition which is natural to check in many cases such that any two processes that are of equal entropy and satisfy the condition must be isomorphic. This has led to the surprising realization that a great many classes of measure-preserving systems that don't seem at all similar to die rolling or coin tossing are in fact isomorphic to independent processes.

The above list spans the core topics of isomorphism theory. In the final chapter of the book, we touch briefly on additional topics in both isomorphism theory and ergodic theory, more broadly construed. In an appendix, we list some of our favorite open problems.

1

Measure-theoretic preliminaries

1. Discussion. In this opening chapter, we offer a review of the basic facts we need from measure theory for the rest of the book (it doubles as an introduction to our pedagogic method). For readers seeking a true introduction to the subject, we recommend first perusing, e.g. Folland (1984); experts meanwhile may safely jump to Chapter 2.

2. Comment. When an exercise is given in the middle of a proof, the end of the exercise will be signaled by a dot: •

The conclusion of a proof is signaled by the box sign at the right margin, thus:

□

1.1. Basic definitions

In this subchapter, we discuss algebras, σ-algebras, generation of a σ-algebra by a family of subsets, completion with respect to a measure and relevant definitions.

3. Definition. Let Ω be a set. An *algebra* of subsets of Ω is a non-empty collection $\mathcal{A}$ of subsets of Ω that is closed under finite unions and complementation. A σ-*algebra* is a collection $\mathcal{A}$ of subsets of Ω that is closed under countable unions and complementation.

4. Comment. Every algebra of subsets of Ω contains the *trivial algebra* $\{\emptyset, \Omega\}$.

5. Exercise. Let Ω be a set and let $\mathcal{C}$ be a family of subsets of Ω. Show that the intersection of all σ-algebras of subsets of Ω containing $\mathcal{C}$ is itself a σ-algebra.

6. Definition. Let Ω be a set and let $\mathcal{C}$ be a family of subsets of Ω. The σ-*algebra generated by* $\mathcal{C}$ is the intersection of all σ-algebras of subsets of Ω containing $\mathcal{C}$.

7. Definition. Let $\mathcal{A}$ be an algebra of subsets of Ω. A *premeasure* on $\mathcal{A}$ is a finitely additive set function p taking $\mathcal{A}$ to the non-negative reals that also never violates countable additivity except for "undefined" cases caused by $\mathcal{A}$ not being a σ-algebra.[2]

If $\mathcal{A}$ is a σ-algebra, then p is called a *measure*.

8. Definition. Let Ω be a set, and let $\mathcal{A}$ be a σ-algebra of subsets of Ω. The pair $(\Omega, \mathcal{A})$ is called a *measurable space*, and the members of $\mathcal{A}$ are called *measurable sets*. Next let μ be a probability measure defined on $\mathcal{A}$; that is, a measure satisfying $\mu(\Omega) = 1$. The triple $(\Omega, \mathcal{A}, \mu)$ is called a *probability space*.

Let $(\Omega, \mathcal{A}, \mu)$ be a probability space. An *event* is a measurable set, that is, a member of $\mathcal{A}$. Two events A and B are *independent* if $\mu(A \cap B) = \mu(A)\mu(B)$.

Let $(\Omega, \mathcal{A}, \mu)$ be a probability space. A *null set* is a set $A \in \mathcal{A}$ with $\mu(A) = 0$. $\mathcal{A}$ is said to be *complete with respect to* μ, or $(\Omega, \mathcal{A}, \mu)$ is simply said to be *complete*, if all subsets of null sets are measurable (and hence null sets).

9. Exercise. Let $(\Omega, \mathcal{B}, \mu)$ be a probability space and suppose that $\mathcal{B}$ is not complete with respect to μ. Let $\mathcal{A} = \{B \cup C : B \in \mathcal{B}, \text{there exists a null set D with } C \subset D\}$. Extend μ to $\mathcal{A}$ by the rule $\mu(B \cup C) = \mu(B)$ for the relevant cases. Show that this extension is well defined and that $(\Omega, \mathcal{A}, \mu)$ is a complete probability space.

10. Definition. The *completion* of a probability space $(\Omega, \mathcal{B}, \mu)$ is the probability space $(\Omega, \mathcal{A}, \mu)$ constructed in the previous exercise.

11. Definition. If A and B are sets, the *symmetric difference* of A and B is the set of points in A or B but not both. We denote the symmetric difference by $A \triangle B$.

12. Definition. Suppose that $(\Omega, \mathcal{A}, \mu)$ is a complete measure space and suppose that $\mathcal{C} \subset \mathcal{A}$ is a family of measurable sets. We say that $\mathcal{C}$ *generates* $\mathcal{A}$ *mod zero* if for every $A \in \mathcal{A}$ there exists B in the σ-algebra $\mathcal{B}$ generated by $\mathcal{C}$ such that $\mu(A \triangle B) = 0$.

13. Comment. Notice that this does not imply that $\mathcal{B} = \mathcal{A}$.

14. Definition. Let Ω be a set and denote its power set by $\mathcal{P}(\Omega)$. An *outer measure* on Ω is a non-increasing, countably sub-additive set function μ^* from $\mathcal{P}(\Omega)$ to the non-negative reals taking the empty set to zero.[3]

---

[2] That is to say, $p : \mathcal{A} \to [0, \infty]$ and if $(A_i)_{i=1}^{\infty} \subset \mathcal{A}$ is pairwise disjoint with $A = \bigcup_{i=1}^{\infty} A_i \in \mathcal{A}$ then $p(A) = \sum_{i=1}^{\infty} p(A_i)$. Notice that $p(\emptyset) = 0$.

[3] That is, $\mu^* : \mathcal{P}(\Omega) \to [0, \infty]$ with $\mu^*(\emptyset) = 0$, $\mu^*(A) \le \mu^*(B)$ whenever $A \subset B$, and $\mu^*(\bigcup_{i=1}^{\infty} A_i) \le \sum_{i=1}^{\infty} \mu^*(A_i)$ for any sequence $(A_i)_{i=1}^{\infty} \subset \mathcal{P}(\Omega)$.

1.2. Carathéodory's theorem, isomorphism, Lebesgue spaces

In this subchapter we develop the machinery for constructing probability spaces and determining when they are essentially the same. We use this machinery to construct Lebesgue measure on the unit interval and define Lebesgue spaces to be those spaces isomorphic to an interval space.

15. Convention. If $(x_n)_{n=1}^{\infty}$ is a sequence, we use the notation $(x_n) \subset X$ to relate the fact that $x_n \in X$ for all n.

16. Theorem. *(Carathéodory; see e.g. Folland 1984, Theorem 1.11.) Let Ω be a set, $\mathcal{A}$ an algebra of subsets of Ω and p a premeasure on $\mathcal{A}$ for which $p(\Omega) = 1$. For every $B \subset \Omega$ let*

$$\mu^*(B) = \inf \left\{ \sum_{i=1}^{\infty} p(A_i) : (A_i)_{i=1}^{\infty} \subset \mathcal{A}, \ B \subset \bigcup_{i=1}^{\infty} A_i \right\}.$$

Let $\mathcal{B} = \{B \subset \Omega : \mu^*(B) + \mu^*(B^c) = 1\}$. Then μ^* is an outer measure on Ω which agrees with p on $\mathcal{A}$, $\mathcal{B}$ is a σ-algebra containing $\mathcal{A}$, and the restriction μ of μ^* to $\mathcal{B}$ is a measure.

(Proof omitted.)

17. Exercise. Show that the measure space arrived at in an application of Carathéodory's theorem is complete.

We now give a couple of applications of Carathéodory's theorem.

18. Definition. Let Λ be a countable set and let $\Omega = \Lambda^{\mathbf{Z}}$. A *cylinder set* is a subset of Ω you get by specifying values for finitely many (possibly zero) coordinates.[4] The *support* of a cylinder set is the set of coordinates whose values are specified.[5]

19. Example. The set of $(x_i)_{i=-\infty}^{\infty} \in \Omega$ such that $x_0 = a$, $x_{17} = b$ and $x_{-2} = c$ is a cylinder set. The support of this cylinder set is $\{-2, 0, 17\}$.

20. Theorem. *(Tychonoff; see e.g. Folland 1984, Theorem 4.43.) Let X_i be compact topological spaces, $i \in \mathcal{I}$, and let $X = \prod_{i \in \mathcal{I}}$. Then X is compact in the product topology.*

21. Exercise. Prove Tychonoff's theorem. $\qquad\qquad\qquad\qquad\qquad\qquad\qquad$ □

---

[4] To be more precise: for $r \geq 0$ an integer, $f_1, f_2, \ldots, f_r \in \mathbf{Z}$ and $\lambda_1, \ldots, \lambda_r \subset \Lambda$, set $C = C(f_1, \ldots, f_r, \lambda_1, \ldots, \lambda_r) = \{(x_i)_{i=-\infty}^{\infty} \in \Omega : x_{f_j} = \lambda_j, 1 \leq j \leq r\}$. C is a cylinder set.

[5] So the support of the cylinder set defined in the previous footnote is $\{f_1, f_2, \ldots, f_r\}$.

22. Exercise.

(a) Show that the family $\mathcal{A}$ of unions of finite (possibly empty) collections of cylinder sets in $\Omega = \Lambda^{\mathbf{Z}}$ forms an algebra. Then show that if Λ is finite and $(A_i)_{i=1}^{\infty}$ are pairwise disjoint members of $\mathcal{A}$ whose union is a member of $\mathcal{A}$ then only finitely many of the A_i are non-empty. *Hint: Apply Tychonoff's theorem to $\Lambda^{\mathbf{Z}}$.*

(b) (See, e.g. McCutchen 1999, Theorem 3.2.4.) If Λ is finite, any finitely additive set function p from cylinder sets to the non-negative reals[6] taking Ω to 1 is extendable to a premeasure on $\mathcal{A}$ which may thereby be extended to a measure by Carathéodory's theorem.[7]

23. Exercise. Let $\Omega = [0, 1)$ and denote by $\mathcal{A}$ the set of unions of finite (possibly empty), pairwise disjoint families of half-open intervals $[a, b) \subset [0, 1)$. Show that $\mathcal{A}$ is an algebra of sets. Put $p\big([a, b)\big) = b - a$. Show that p has a unique extension to a premeasure on $\mathcal{A}$.

24. Definition. The outer measure μ^* you get by applying Carathéodory's theorem to the premeasure p of the foregoing exercise is called *Lebesgue outer measure,* which we denote by m^*. We denote by $\mathcal{L}$ the σ-algebra $\mathcal{B}$ coming from Carathéodory's theorem; members of $\mathcal{L}$ are called *Lebesgue measurable sets.* The restriction of m^* to $\mathcal{B}$ is called *Lebesgue measure,* which we denote by m.

25. Remark. Although we've only defined Lebesgue measure, Lebesgue measurable sets, etc. here on the unit interval, one can of course extend this to the whole line in the obvious way; readers should convince themselves of this.

26. Definition. For $A \subset \mathbf{R}$, define the *Lebesgue inner measure* of A to be the quantity $m_*(A) = \sup\{m(B) : B \in \mathcal{L}, B \subset S\}$.[8]

27. Exercise. Prove that for any set A and any interval I, $m_*(A) = |I| - m^*(I \setminus A)$.

---

[6] In other words, if $C_1, \ldots, C_r$ are pairwise disjoint cylinder sets whose union is a cylinder set C, then $p(C) = \sum_{i=1}^{r} p(C_i)$.

[7] It is instructive, and the reader is encouraged, to explore just what such a finitely additive function looks like. For a quick example, suppose that $\Lambda = \{a, d\}$. The cylinder set C_1 that sees the occurrence of "add" at the zero place (that is, $C(0, 1, 2, a, d, d)$) and the cylinder set C_2 that sees "ada" at the zero place are disjoint and their union is the cylinder set C_3 that sees "ad" at the zero place; hence any premeasure p must satisfy $p(C_1) + p(C_2) = p(C_3)$, but these are the only sorts of conditions.

[8] The reader should check that this supremum is a maximum; i.e. it is attained.

28. Exercise. Show that for any Lebesgue measurable set $A \subset [0, 1]$, $m(A) = \sup\{m(K) : K \subset A, K \text{ is closed}\}$ and $m(A) = \inf\{m(U) : A \subset U, U \text{ is open}\}$. (We may sometimes say, accordingly, that Lebesgue measure is *inner regular* with respect to closed sets and *outer regular* with respect to open sets.)

29. Definition. Let $(\Omega, \mathcal{A})$ and $(\Omega', \mathcal{A}')$ be measurable spaces. A function $T : \Omega \to \Omega'$ satisfying $T^{-1}A' \in \mathcal{A}$ for every $A' \in \mathcal{A}'$ is said to be $(\mathcal{A}, \mathcal{A}')$-*measurable,* or simply *measurable* when $\mathcal{A}$ and $\mathcal{A}'$ are understood. Let $(\Omega, \mathcal{A}, \mu)$ and $(\Omega', \mathcal{A}', \mu')$ be probability spaces. A measurable function $T : \Omega \to \Omega'$ satisfying $\mu(T^{-1}A') = \mu'(A')$ for every $A' \in \mathcal{A}'$ is said to be *measure-preserving.*

30. Theorem. *(Urysohn's lemma; see e.g. Dudley 2002, Lemma 2.6.3.) Let X be a normal topological space, and let A and B be disjoint closed subsets of X. There exists a continuous function $f : X \to [0, 1]$ such that $f(x) = 0$ for all $x \in A$ and $f(x) = 1$ for all $x \in B$.*

(Proof omitted.)

31. Exercise. Let $(\Omega, \mathcal{A}, \mu)$ and $(\Omega', \mathcal{A}', \mu')$ be measure spaces and suppose that $\mathcal{A}'$ is generated by a family of sets $\mathcal{B}$. Let $T : \Omega \to \Omega'$. Show that:

(a) if $T^{-1}B \in \mathcal{A}$ for every $B \in \mathcal{B}$ then T is measurable;
(b) if $\mu(T^{-1}B) = \mu'(B)$ for every $B \in \mathcal{B}$ then T is measure-preserving.

32. Definition. Let $(\Omega, \mathcal{A}, \mu)$ and $(\Omega', \mathcal{A}', \mu')$ be probability spaces and suppose $\pi : \Omega \to \Omega'$ is a measure-preserving transformation. We say that π is a *homomorphism,* or a *factor map,* and that $(\Omega', \mathcal{A}', \mu')$ is a *factor* of $(\Omega, \mathcal{A}, \mu)$.

33. Definition. Let $(\Omega, \mathcal{A}, \mu)$ and $(\Omega', \mathcal{A}', \mu')$ be probability spaces and suppose $T : \Omega \to \Omega'$ is a homomorphism. If there exist full measure sets[9] $X \subset \Omega$ and $X' \subset \Omega'$ such that the restriction of T to X is a bijection to X', and $T^{-1} : X' \to X$ is measurable, then we will say that T is an *isomorphism,* and that the spaces $(\Omega, \mathcal{A}, \mu)$ and $(\Omega', \mathcal{A}', \mu')$ are *isomorphic.*

34. Comment. When two spaces $(\Omega, \mathcal{A}, \mu)$ and $(\Omega', \mathcal{A}', \mu')$ are isomorphic, then, once appropriate null sets are disregarded, they are "essentially the same space". In other words, they are in fact the same, up to relabeling.

35. Exercise. Show that "is isomorphic to" is an equivalence relation and that completeness is an isomorphism invariant.

---

[9] A set $X \subset \Omega$ is said to be of full measure if $\mu(\Omega \setminus X) = 0$.

36. Convention. Our attitude toward null sets is that they "don't count". Accordingly we will assume that all probability spaces are complete.

Without this convention, $([0, 1], \mathcal{L}, m)$ would not be isomorphic to $([0, 1], \mathcal{B}, m)$, where $\mathcal{B}$ is the σ-algebra of Borel sets (that is, the σ-algebra generated by the open sets) for the rather uninteresting (though non-trivial) reason that $\mathcal{L}$ has more null sets than $\mathcal{B}$ does.

37. Definition. An *interval space* consists of an interval $[0, t]$ equipped with Lebesgue measure, where $0 \le t \le 1$, to which are appended countably many points having a combined positive measure $1 - t$; *with atoms if $t < 1$, without atoms if $t = 1$.*[10]

38. Exercise. Show that an interval space is a complete probability space.

39. Definition. A *Lebesgue space* is a probability space that is isomorphic to some interval space. A Lebesgue space is *non-atomic* if it is isomorphic to $([0, 1], \mathcal{L}, m)$.

40. Remark. A classic reference in the theory of Lebesgue spaces, axiomatically defined, is Rohlin (1952), though this and most of the literature on axiomatic treatments is fraught with vagueness and ambiguity (if not confusion), due in part to a cavalier attitude toward sets of measure zero. (For an interesting and well-motivated modern axiomatic treatment, see Rudolph (1990).) An arguably more sensible theory of spaces measurably isomorphic to the unit interval is that of regular Borel probability measures on Polish spaces; however, we are choosing to skirt most of the issues entirely by simply defining Lebesgue spaces to be those that are isomorphic to an interval space. (Not all of the issues: see the axiomatic criterion in Theorem 53 below.)

There are tremendous technical advantages to doing analysis on Lebesgue spaces. Following standard ergodic theory practice, we shall deal almost exclusively with non-atomic Lebesgue spaces in this book. Since, in essence, the only non-atomic Lebesgue space is the unit interval, one can be deceived into thinking that this is unduly restrictive. However, the concept is actually quite general. Indeed, just about every probability space you are likely to encounter is Lebesgue or at least has a Lebesgue completion; in particular, spaces derived from completing regular Borel measures on compact metrizable spaces are Lebesgue; non-Lebesgue spaces are pathological examples, generally deriving

---

[10] So, let $\theta = [0, t] \cup C$, where C is a countable set whose intersection with $[0, t]$ is empty, let $f : C \to [0, 1 - t]$ be a function satisfying $\sum_{c \in C} f(c) = 1 - t$, let $\mathcal{A}$ consist of all $L \cup D$, where L is a Lebesgue measurable subset of $[0, t]$ and D is any subset of C, and for $A = L \cup D \in \mathcal{A}$, put $\mu(A) = m(L) + \sum_{c \in D} f(c)$. $(\theta, \mathcal{A}, \mu)$ is an interval space.

from non-separable topologies or via constructions requiring the axiom of choice.

1.3. Properties of Lebesgue spaces, factors

In this subchapter we collect facts about probability spaces for which the hypothesis that you are dealing with a Lebesgue space is needed. Among these is the fact that any factor of a Lebesgue space is a Lebesgue space.

41. Exercise. Let $(\Omega, \mathcal{A}, \mu)$ be a non-atomic Lebesgue space. Let $\epsilon > 0$ and suppose $A \in \mathcal{A}$ with $\mu(A) > \epsilon$. Then there exists $B \in \mathcal{A}$ with $B \subset A$ satisfying $\mu(B) = \epsilon$.

We need the following classical theorem.

42. Theorem. *(Lusin; see e.g. Folland 1984, Theorem 7.10.) Let $f : [0, 1] \to \mathbf{R}$ be Lebesgue measurable. For every $\epsilon > 0$ there exists a continuous function $g : [0, 1] \to \mathbf{R}$ such that $m(\{x : f(x) \neq g(x)\}) < \epsilon$.*

Sketch of proof.

43. Exercise. Show that we may assume without loss of generality that $0 < f(x) < 1$ for all $x \in [0, 1]$. *Hint: Consider the function* $h(x) = \frac{1}{2}\left(1 + \frac{f(x)}{1+\sup(|f|)}\right)$. •

Suppose, then, that $0 < f(x) < 1$. For every $x \in [0, 1]$, write $f(x) = \sum_{n=1}^{\infty} a_n(x)2^{-n}$ for the binary decimal expansion of $f(x)$[11] and let $H_n = \{x \in [0, 1] : a_n(x) = 1\}$. The sets H_n are clearly Lebesgue measurable. Accordingly, by Exercise 28 one may approximate H_n by an open set U_n and by a closed set K_n such that $K_n \subset H_n \subset U_n$, with $m(U_n \setminus K_n) < \delta_n$, where $\sum_{n=1}^{\infty} \delta_n < \epsilon$. Pick, by Urysohn's lemma, continuous functions h_n satisfying $h_n(x) = 1$ for $x \in K_n$ and $h_n(x) = 0$ for $x \in U_n^c$. Let $g(x) = \sum_{n=1}^{\infty} h_n(x)2^{-n}$.

44. Exercise. Show that g is continuous and that $m(\{x : f(x) \neq g(x)\}) < \epsilon$. □

Below, we will see that a measure-preserving bijection T between probability spaces needn't be an isomorphism (due to non-measurability of T^{-1}) even when the range space is Lebesgue. The content of the next theorem is that this can't happen when the domain and range are each Lebesgue.

---

[11] There is an issue about uniqueness here for countably many cases but one can agree to choose in these cases in some canonical fashion; for example, one can agree to avoid expansions having only finitely many occurrences of 0 where possible.

45. Theorem. *Let $(\Omega, \mathcal{A}, \mu)$ and $(X, \mathcal{B}, \nu)$ be Lebesgue spaces and suppose $T : \Omega \to X$ is injective and measure preserving. Then TA is measurable for every measurable set A.*

Sketch of proof.

46. Exercise. Prove that it is sufficient for the general case to establish the conclusion when both $(\Omega, \mathcal{A}, \mu)$ and $(X, \mathcal{B}, \nu)$ are $([0, 1], \mathcal{L}, m)$.   •

So, let $T : [0, 1] \to [0, 1]$ be injective and measure-preserving. Let $A \in \mathcal{L}$. We must show that $TA \in \mathcal{L}$.

47. Exercise. It is sufficient to show that for arbitrary $\epsilon > 0$, $m^*(TA) < m_*(TA) + 2\epsilon$.   •

Let $\epsilon > 0$ be arbitrary. By Lusin's theorem, there is a continuous function $g : [0, 1] \to [0, 1]$ such that the set $B = \{x : Tx \neq g(x)\}$ satisfies $m(B) < \epsilon$. ϵ-approximate $A \setminus B$ and $A^c \setminus B$ from the inside by closed sets K_1 and K_2, respectively.

48. Exercise. Show that $g(K_1)$ and $g(K_2)$ are measurable. *Hint: recall that the continuous image of a compact set is compact.* Show that $m(K_1) + m(K_2) > 1 - 2\epsilon$. Finally show that $g(K_1) \subset TA \subset g(K_2)^c$ and complete the proof. □

49. Exercise. Obtain the following more general version of Theorem 45 by adapting its proof. Let $(\Omega, \mathcal{A}, \mu)$ and $(X, \mathcal{B}, \nu)$ be Lebesgue spaces and suppose $T : \Omega \to X$ is measure-preserving. Prove that if for some $B \subset X$, $T^{-1}(B)$ is measurable, then B is measurable.

We need a way of verifying that spaces we construct are Lebesgue spaces.

50. Definition. Let Ω be a set. A sequence $(C_i)_{i=1}^{\infty}$ of subsets of Ω is said to *separate points* on Ω if for every $x, y \in \Omega$ with $x \neq y$ there is some $i \in \mathbf{N}$ such that either $x \in C_i$ and $y \in C_i^c$ or $y \in C_i$ and $x \in C_i^c$. A sequence $(P_i)_{i=1}^{\infty}$ of countable partitions of Ω separates points if a sequence consisting of all the cells of all the P_i separates points.

51. Exercise. Let $(\Omega, \mathcal{A}, \mu)$ be a measure space. Show that $\rho(A, B) = \mu(A \triangle B)$ defines a metric on $\mathcal{A}$, modulo identification of sets differing by a null set.

52. Definition. Let $(\Omega, \mathcal{A}, \mu)$ be a measure space. A property P predicated of points $\omega \in \Omega$ is said to hold *almost everywhere*, or *a.e.*, if the set of x for which $P(x)$ fails has zero measure.

53. Theorem. *Let $(\Omega, \mathcal{A}, \mu)$ be a probability space. Suppose there is a countable sequence of measurable sets $(C_i)_{i=1}^{\infty}$ that separates points such that:*

(a) *if B_i is a sequence of measurable sets such that $B_i \in \{C_i, C_i^c\}$ for every $i \in \mathbf{N}$ then $\bigcap_{i=1}^{\infty} B_i \neq \emptyset$;*

(b) *if C is the algebra generated by $(C_i)_{i=1}^{\infty}$ then for every $A \in \mathcal{A}$, $\mu(A) = \inf\{\sum_{i=1}^{\infty} \mu(D_i) : D_i \in C, i \in \mathbf{N} \text{ with } A \subset \bigcup_{i=1}^{\infty} D_i\}$.*

Then $(\Omega, \mathcal{A}, \mu)$ *is a Lebesgue space.*

Idea of proof. First establish that there is a countable sequence $(B_i)_{i=1}^{\infty} \subset \mathcal{A}$ that is dense for the metric $\rho(A, B) = \mu(A \triangle B)$. Next, let P_n be the algebra generated by $B_1, B_1^c, \ldots, B_n, B_n^c$. (P_n) is an increasing sequence of partitions. Construct an isomorphic increasing sequence of partitions of the unit interval that separates points and use this to construct a pointwise isomorphism.

Sketch of proof. First we assume that $\mu(\{\omega\}) = 0$ for every $\omega \in \Omega$. That is, we assume that $(\Omega, \mathcal{A}, \mu)$ has no atoms. For $n \in \mathbf{N}$, let C_n be the set of positive measure, minimal elements of the algebra generated by $(C_i)_{i=1}^{n}$. Let $\mathcal{I}$ be the family of closed intervals contained in $[0, 1]$.

54. Exercise. Show there exist, for all $n \in \mathbf{N}$, maps $\Phi_n : C_n \to \mathcal{I}$ satisfying:

(a) for every $n \in \mathbf{N}$ and every $C \in C_n$, $m\big(\Phi_n(C)\big) = \mu(C)$;

(b) for every $n \in \mathbf{N}$ and every distinct $C, D \in C_n$, $\Phi_n(C) \cap \Phi_n(D)$ is either empty or consists of a single point;

(c) if $n > m$, $C \in C_n$ and $D \in C_m$ with $C \subset D$, then $\Phi_n(C) \subset \Phi_m(D)$.

55. Exercise.

(a) Show that for a.e. $\omega \in \Omega$, ω lies in some member $C_{n,\omega}$ of C_n for every $n \in \mathbf{N}$.

(b) For such ω, prove that $\lim_n \mu(C_{n,\omega}) = 0$. Use this to show that $\bigcap_{n=1}^{\infty} \Phi_n(C_{n,\omega})$ has a unique element $\pi(\omega)$.

(c) Prove that $\pi : \Omega \to [0, 1]$ is, upon deleting suitable sets of measure zero from Ω and from $[0, 1]$, a bimeasurable measure-preserving bijection, i.e. an isomorphism. *Hint: measurability of the inverse is achieved by, essentially, showing the image to be full measure.* ●

For the general case, when $(\Omega, \mathcal{A}, \mu)$ may have atoms, start by deleting the atoms and applying the first case to show that whatever part of the space remains is isomorphic to some interval $[0, t]$. The rest is easy: just map atoms to atoms of equal measure. □

We now turn to a discussion of how a sub-σ-algebra of a Lebesgue space determines a factor of the whole space that is itself Lebesgue.

56. Example. Notice that if $\pi : \Omega \to \Omega'$ is a factor map then $\pi^{-1}(\mathcal{A}') = \{\pi^{-1}(A) : A \in \mathcal{A}'\}$ is a σ-algebra contained in $\mathcal{A}$. The following theorem contains a converse.

57. Theorem. *Let $(\Omega, \mathcal{A}, \mu)$ be a Lebesgue probability space and suppose that $\mathcal{B} \subset \mathcal{A}$ is a σ-algebra. There exists a Lebesgue space $(\Omega', \mathcal{A}', \mu')$ and a factor map $\pi : \Omega \to \Omega'$ such that $\pi^{-1}(\mathcal{A}') = \mathcal{B}$.*

Idea of proof. Choose a countable dense set $(B_i) \subset \mathcal{B}$. Merge points not separated by (B_i). Apply Carathéodory.

Sketch of proof. Let (B_i) be as above.

58. Exercise. We can assume without loss of generality that the members of this sequence form an algebra of sets. ●

For $x, y \in \Omega$, write $x \sim y$ if for all $i \in \mathbf{N}$, x and y are in the same cell of the partition $\{B_i, B_i^c\}$.

59. Exercise. Show that $\sim$ is an equivalence relation. ●

Let Ω' be the set of equivalence classes of $\sim$ and let $\pi : \Omega \to \Omega'$ be the map that sends a point ω to its equivalence class. (If $x \in \Omega$ then $\pi^{-1}\big(\pi(x)\big)$, that is, the equivalence class of x under $\sim$, is called *the fiber over x*.)

60. Exercise. $\big(\pi(B_i)\big)_{i=1}^{\infty}$ forms an algebra of subsets of Ω'. If we let $p\big(\pi(B_i)\big) = \mu(B_i)$, $i \in \mathbf{N}$, p is a premeasure on this algebra. ●

Using Carathéodory's extension theorem and the standard completeness construction, extend the algebra to a σ-algebra $\mathcal{A}'$ and the premeasure p to a complete measure μ'.

61. Exercise. Show that $\pi : \Omega \to \Omega'$ is a homomorphism from $(\Omega, \mathcal{A}, \mu)$ to $(\Omega', \mathcal{A}', \mu')$ satisfying $\pi^{-1}(\mathcal{A}') = \mathcal{B}$. Next show that $(\Omega', \mathcal{A}', \mu')$ is a Lebesgue space. *Hint: it may have atoms even if $(\Omega, \mathcal{A}, \mu)$ doesn't. Get rid of these first and apply Theorem 53.* □

62. Convention. If $(\Omega, \mathcal{A}, \mu)$ is a Lebesgue space and $\mathcal{B} \subset \mathcal{A}$ is a σ-algebra, we may say that $\mathcal{B}$ is a "factor" of $\mathcal{A}$. What we have in mind by this abuse of notation is the space $(\Omega', \mathcal{A}', \mu')$ of the above construction.

63. Lemma. *Let $(\Omega, \mathcal{A}, \mu)$ be a probability space and let $f : \Omega \to \mathbf{N}$ be a measurable function. Given ϵ, there is an i such that $P(f \geq i) < \epsilon$.*

64. Exercise. Prove the lemma. □

65. Definition. Let $(\Omega, \mathcal{A}, \mu)$ be a probability space. A *measurable partition* is a partition P of Ω into measurable cells.

66. Theorem. *Suppose $(\Omega, \mathcal{A}, \mu)$ is a Lebesgue space. Let $(P_i)_{i=1}^{\infty}$ be an increasing[12] sequence of finite, measurable partitions that separates points. If $A \in \mathcal{A}$ and $\epsilon > 0$, there is some $i \in \mathbf{N}$ and some set B which is a union of some of the pieces of P_i such that $\mu(A \triangle B) < \epsilon$.*

Idea of proof. Assume your space to be $[0, 1)$. Remove a little of the space to make A, A^c, and every set of P_1 closed. Remove a little more to make every set of P_2 closed. Continue. Since nested sequences of these closed sets intersect in at most a single point, they are eventually either completely out of A or completely out of A^c. For a point x in your space, let $i(x)$ be the smallest natural number so that the atom of P_i that contains x is either completely out of A or completely out of A^c. Then most points of P_i are in an atom completely out of A or completely out of A^c for i chosen as in the previous lemma.

Sketch of proof.

67. Exercise. Show that the conclusion of the theorem is an isomorphism invariant, so that in particular we may assume without loss of generality that $(\Omega, \mathcal{A}, \mu)$ is an interval space. Indeed, show that we may assume that $(\Omega, \mathcal{A}, \mu)$ is an interval space without atoms; i.e. $[0, 1]$ with Lebesgue measure. •

A big advantage of being allowed to assume the space is $([0, 1], \mathcal{L}, m)$ is that we can utilize the relationship of m to the topology of $[0, 1]$; in particular, the facts that the measure m is inner regular with regard to compact sets and outer regular with regard to open sets. Let $A \in \mathcal{A}$ and let $\epsilon > 0$.

68. Exercise. Show that the cells comprising the partitions P_i can be approximated by closed sets from the inside in such a way that:

(a) for all n, the union of the closed sets approximating the cells of P_n has measure $> 1 - \frac{\epsilon}{2}$;
(b) for any sequence $(A_i)_{i=1}^{\infty}$ satisfying $A_i \in P_i$ and $A_1 \supset A_2 \supset A_3 \supset \cdots$, one has $C_1 \supset C_2 \supset C_3 \supset \cdots$, where C_i is the closed set approximating A_i. •

69. Exercise. Show that for every $\omega \in \Omega$, there exist sets $A_i^{(\omega)} \in P_i$, $i \in \mathbf{N}$, such that $\bigcap_{i=1}^{n} A_i^{(\omega)} = \{\omega\}$. It follows that if $C_i^{(\omega)}$ is the interior approximating

---

[12] By increasing we mean if $i < j$ and $p \in P_i$ then p is a union of some members of P_j.

closed set for $A_i^{(\omega)}$, one has $\bigcap_{i=1}^{n} C_i^{(\omega)}$ equal to either $\{\omega\}$ or to $\emptyset$. Show that the measure of $P = \{\omega : \bigcap_{i=1}^{n} C_i^{(\omega)} = \{\omega\}\}$ is at least $1 - \frac{\epsilon}{2}$. •

Now, let C be an open set containing A and satisfying $m(C) < m(A) + \frac{\epsilon}{2}$.

70. Exercise. For any $x \in P \cap C$, there exists $r = r(x) \in \mathbf{N}$ such that $C_r^{(\omega)} \subset C$. Hence for some $R \in \mathbf{N}$, $E = \{x \in P \cap C : r(x) \leq R\}$ satisfies $m(E) > m(A) - \frac{\epsilon}{2}$. •

One finishes the proof by letting $B = \bigcup_{\omega \in E} C_R^{(\omega)}$. □

71. Corollary. *Let $\mathcal{P}$ be the family of measurable sets consisting of the cells of the various partitions P_i in Theorem 66. Then $\mathcal{P}$ generates $\mathcal{A}$ mod 0.*

1.4. Random variables, integration, (stationary) processes

In this subchapter we collect the basic facts of integration theory and introduce one of the most important objects of study for this book: stationary processes.

72. Definition. By a *random variable* we mean a measurable function from a probability space to an abstract set Λ equipped with a σ-algebra. We will call Λ an *alphabet*.

73. Discussion. Most of the random variables we encounter in this book will be into a finite alphabet. For real-valued random variables, we require a bit of integration theory, which we now develop briefly.

Let $(\Omega, \mathcal{A}, \mu)$ be a fixed measure space.

74. Definition. The *extended real line* is the set $\overline{\mathbf{R}} = \mathbf{R} \cup \{-\infty, +\infty\}$. The usual order relation $<$ is extended to a total order on $\overline{\mathbf{R}} \times \overline{\mathbf{R}}$ consistent with $-\infty < x < +\infty$ for all real x.

75. Notation. Notation for intervals in $\overline{\mathbf{R}}$ is exactly what one would expect, consistent with the extended order relation, e.g. $[-\infty, x) = \{y \in \overline{\mathbf{R}} : y < x\}$.

76. Definition. The *order topology* on $\overline{\mathbf{R}}$ is the topology generated by intervals; in particular, intervals $[-\infty, x)$ form a neighborhood basis for $-\infty$, while intervals $(x, +\infty]$ form a neighborhood basis for $+\infty$.

77. Convention. When discussing measurability of functions into the extended reals, we mean with respect to the σ-algebra on $\overline{\mathbf{R}}$ generated by the open sets of the order topology.

78. Exercise. If $(f_i)_{i=1}^{\infty}$ are measurable functions $\Omega \to \overline{\mathbf{R}}$ then $\sup_i f_i$, $\inf_i f_i$, $\limsup_i f_i$ and $\liminf_i f_i$ are measurable.

79. Definition. A *simple function* $\Omega \to \mathbf{R}$ is a function $\varphi(x) = \sum_{i=1}^{n} c_i 1_{A_i}(x)$, where $A_1, \ldots, A_n \in \mathcal{A}$ with $\mu(A_i) < \infty$ and $c_1, \ldots, c_n \in \mathbf{R}$. The *integral* of φ is $\int \varphi \, d\mu = \sum_{i=1}^{n} c_i \mu(A_i)$.

80. Exercise. If $f : \Omega \to [0, \infty]$ is measurable, there exists a sequence $(\varphi_i)_{i=1}^{\infty}$ of simple functions with $0 \le \varphi_1(x) < \varphi_2(x) < \cdots$ such that $\lim_{n \to \infty} \varphi_n(x) = f(x)$ for all $x \in \Omega$.

We now extend the definition of the integral to non-negative measurable functions.

81. Definition. If $f : \Omega \to [0, \infty]$ is measurable, let

$$\int f \, d\mu = \sup \left\{ \int \varphi \, d\mu : 0 \le \varphi \le f, \ \varphi \text{ a simple function} \right\}.$$

82. Exercise. This definition of the integral agrees with the previous one if f is a simple function. *Hint: show first that for simple functions $\varphi_1 < \varphi_2$, $\int \varphi_1 \, d\mu < \int \varphi_2 \, d\mu$.*

83. Theorem. *(Monotone convergence theorem.) If $(f_i)_{i=1}^{\infty}$ is a non-decreasing sequence of non-negative measurable functions on Ω and we let $f(x) = \lim_{i \to \infty} f_i(x)$, then $\int f \, d\mu = \lim_{i \to \infty} \int f_i \, d\mu$.*

Sketch of proof. Clearly $\int f_i \, d\mu \le \int f \, d\mu$ for all i, so $\lim_{i \to \infty} \int f_i \, d\mu \le \int f \, d\mu$. Let $\epsilon > 0$ be arbitrary and let $\varphi = \sum_{i=1}^{M} c_i 1_{B_i}$ be an arbitrary non-negative simple function with $\varphi \le f$. The idea is to show that $\lim_{i \to \infty} \int f_i \, d\mu \ge (1 - \epsilon) \int \varphi \, d\mu$, which, since ϵ and φ are arbitrary, will get that $\lim_{i \to \infty} \int f_i \, d\mu \ge \int f \, d\mu$, completing the proof.

84. Exercise. Fill in the details. □

85. Exercise. If $f \to [0, \infty]$ is measurable then $\int f \, d\mu = 0$ if and only if $f(x) = 0$ a.e.

86. Theorem. *(Fatou's lemma.) Let $f_n : \Omega \to [0, \infty]$ be measurable functions, $n \in \mathbf{N}$.*

(a) $\int \liminf_{n \to \infty} f_n \, d\mu \le \liminf_{n \to \infty} \int f_n \, d\mu$.
(b) If $\mu(\Omega) < \infty$ and there exists $K < \infty$ such that $f_n(x) \le K$ for all x and all n then $\int \limsup_{n \to \infty} f_n \, d\mu \ge \limsup_{n \to \infty} \int f_n \, d\mu$.

Sketch of proof.

(a) For all i, $\int \inf_{n \ge i} f_n \, d\mu \le \inf_{n \ge i} \int f_n \, d\mu$. Notice that $\inf_{n \ge i} f_n$ increases to $\liminf_{n \to \infty} f_n$ as $i \to \infty$.

87. Exercise. Complete the proof. *Hint: monotone convergence theorem. For (b), let $g_n = K - f_n$ and apply part (a).*   □

88. Definition. Let $f : \Omega \to \overline{\mathbf{R}}$. We put $f^+ = \sup\{f, 0\}$ and $f^- = -\inf\{f, 0\}$.

89. Definition. If $f : \Omega \to [-\infty, \infty]$ is measurable and at least one of $\int f^+ \, d\mu$, $\int f^- \, d\mu$ is finite then we let $\int f \, d\mu = \int f^+ \, d\mu - \int f^- \, d\mu$. If $\int f \, d\mu$ exists and is finite we say that f is *integrable*.

90. Comment. One easily checks that f is integrable if and only if $\int |f| \, d\mu < \infty$.

91. Exercise. If $f(x) = g(x)$ a.e. then $\int f \, d\mu = \int g \, d\mu$.

92. Theorem. *(Dominated convergence theorem.) Suppose that $(f_i)_{i=1}^\infty$ is a sequence of measurable functions $\Omega \to \mathbf{R}$ and $g : \Omega \to [0, +\infty]$ with $|f_i| \le g$ a.e. for all i. If $f(x) = \lim_{n\to\infty} f_n(x)$ exists a.e. then f is integrable and $\int f \, d\mu = \lim_{n\to\infty} \int f_n \, d\mu$.*

Sketch of proof. Note that $g - f_n \ge 0$ a.e. By Fatou's lemma,

$$\int g \, d\mu - \int f \, d\mu = \int \liminf_{n\to\infty}(g - f_n) \, d\mu \le \liminf_{n\to\infty} \int g - f_n \, d\mu$$
$$= \int g \, d\mu - \limsup_{n\to\infty} \int f_n \, d\mu.$$

93. Exercise. Complete the proof. *Hint: rerun the above for $g + f_n$; conclude that* $\limsup_{n\to\infty} \int f_n \, d\mu \le \int f \, d\mu \le \liminf_{n\to\infty} \int f_n \, d\mu$.   □

94. Definition. If X is an integrable random variable into the reals, the *expectation* of X is defined by $E(X) = \int X$.

95. Convention. In this book, random variables will, unless otherwise noted, be functions into countable alphabets Λ, and the relevant σ-algebra on Λ will always be taken to be the power set $\mathcal{P}(\Lambda)$. We will often say simply "X is a random variable on the alphabet Λ", suppressing mention of the measure space $(Z, \mathcal{B}, \mu)$ on which X is defined. Note, however, that when multiple random variables are considered simultaneously, we always assume that their domain spaces coincide (see below for more detailed conventions).

96. Notation. If X is a random variable from domain space $(Z, \mathcal{B}, \mu)$ to an alphabet Λ, for a set $B \subset \Lambda$ we write $P(X \in B)$ (read "the probability that X lies in B") for the quantity $\mu(X^{-1}(B))$. If $B = \{\lambda\}$ is a singleton, we may write $P(X = \lambda)$ (read "the probability that X is equal to λ") instead. Also, for example, if X_1, X_2, X_3 are random variables we write $P(X_1 = a, X_2 = b, X_3 = c)$ for $\mu(X_1^{-1}(a) \cap X_2^{-1}(b) \cap X_3^{-1}(c))$.[13]

97. Definition. The *density function* of a random variable X on an alphabet Λ is the function $f : \Lambda \to [0, 1]$ defined by $f(\lambda) = P(X = \lambda)$. The *joint density function* of random variables X_1, X_2, X_3 is given by $g(a, b, c) = P(X_1 = a, X_2 = b, X_3 = c)$.[14]

98. Definition. Random variables $X_1, X_2, \ldots, X_r$ are said to be *independent* if their joint density function is the product of their individual density functions.[15] An infinite family of random variables is independent if every finite subfamily is independent.

99. Definition. Let Λ be an alphabet. A *preprocess* on Λ is a doubly infinite[16] sequence of random variables $\ldots, X_{-3}, X_{-2}, X_{-1}, X_0, X_1, X_2, X_3, \ldots$ into an alphabet Λ. It is *stationary* if its "probability law" is "time invariant".[17]

100. Definition. A preprocess $(X_i)_{i=-\infty}^{\infty}$ on Λ is a *process* if its domain space (that is, the domain space of the X_i) is $(\Omega, \mathcal{A}, \mu)$, where $\Omega = \Lambda^{\mathbf{Z}}$, $\mathcal{A}$ is generated by the cylinder sets,

$$P(X_{i_1} = \lambda_1, \ldots, X_{i_k} = \lambda_k) = \mu\{(x_i)_{i=-\infty}^{\infty} \in \Omega : x_{i_1} = \lambda_1, \ldots, x_{i_k} = \lambda_k\}$$

for all $i_1, \ldots, i_k \in \mathbf{Z}$ and $\lambda_1, \ldots, \lambda_k \in \Lambda$, and $(\Omega, \mathcal{A}, \mu)$ *is a Lebesgue space.*

101. Comment. What we call a preprocess might be called a process by some statisticians. However, this allows in some pathological examples that we would like to avoid; we discuss this below. The crucial part of our definition is the part that requires $(\Omega, \mathcal{A}, \mu)$ to be Lebesgue.

Note that when (X_i) is a stationary process, the probability that X_0 is equal to c, X_1 is equal to a and X_2 is equal to t is the same as the probability that X_{100} is equal to c, X_{101} is equal to a and X_{102} is equal to t. This is just a way of saying that the probability of seeing "cat" doesn't change over time.

---

[13] More generally, if $X_1, \ldots, X_r$ are random variables and $B_1, \ldots, B_r$ are subsets of Λ, we write $P(X_1 \in B_1, \ldots, X_r \in B_r)$ for the quantity $\mu(X^{-1}(B_1) \cap \cdots \cap X^{-1}(B_r))$ or, if $B_i = \{\lambda_i\}$ are singletons, $P(X_1 = \lambda_1, \ldots, X_r = \lambda_r)$.

[14] More generally, if $X_1, \ldots, X_r$ are random variables into Λ, their *joint density function* is the function $g : \Lambda^r \to [0, 1]$ defined by $g(\lambda_1, \ldots, \lambda_r) = P(X_1 = \lambda_1, X_2 = \lambda_2, \ldots, X_r = \lambda_r)$.

[15] In other words, if $P(X_1 = \lambda_1, X_2 = \lambda_2, \ldots, X_r = \lambda_r) = P(X_1 = \lambda_1)P(X_2 = \lambda_2) \ldots P(X_r = \lambda_r)$ for every $\lambda_1, \lambda_2, \ldots, \lambda_r \in \Lambda$.

[16] One can also consider singly infinite sequences, though we usually won't in this book.

[17] This means that for any $f_1, f_2, \ldots, f_r \in \mathbf{Z}$, every $B_1, \ldots, B_r \subset \Lambda$, and every $n \in \mathbf{Z}$, $P(X_{f_i} \in B_i, 1 \le i \le r) = P(X_{f_i + n} \in B_i, 1 \le i \le r)$.

102. Definition. Random variables X and Y are *identically distributed* if their density functions coincide. A stationary process $(X_i)_{i=-\infty}^{\infty}$ is said to be *i.i.d.* (for *independent, identically distributed*) if the X_is are independent and identically distributed.

103. Example. Say we want to represent an independent sequence of heads and tails obtained by a doubly infinite independent sequence of flips of a fair coin. This can be modeled as a stationary process in the following way. Let $\Lambda = \{h, t\}$ and put $\Omega = \Lambda^{\mathbf{Z}}$. For any cylinder set C defined by specifying r coordinates,[18] put $p(C) = 2^{-r}$.

104. Exercise. Show that p extends uniquely to a premeasure on the algebra consisting of the finite unions of cylinder sets. Apply Carathéodory's theorem to extend p to a measure on the σ-algebra generated by the cylinder sets. Denote by $(\Omega, \mathcal{A}, \mu)$ the resulting probability space. Now define a process $(X_i)_{i=-\infty}^{\infty}$ on Ω by the rule $X_j\big((\omega_i)_{i=-\infty}^{\infty}\big) = \omega_j$.

This is an effective model of a doubly infinite independent sequence of flips of a fair coin in the sense that a black box which picks a point $\omega \in \Omega$ "at random" and spits out the sequence $\big(X_i(\omega)\big)_{i=-\infty}^{\infty}$ will not be distinguishable from one that spits out a doubly infinite sequence of hs and ts obtained from coin tossing.

105. Theorem. *Let $(Y_i)_{i=-\infty}^{\infty}$ be any stationary preprocess whose domain space is not necessarily Lebesgue. Show that there exists a stationary process $(X_i)_{i=-\infty}^{\infty}$ such that for any $r \in \mathbf{N}$ and $f_1, \ldots, f_r \in \mathbf{Z}$, the joint distribution of $X_{f_1}, \ldots, X_{f_r}$ is equal to the joint distribution of $Y_{f_1}, \ldots, Y_{f_r}$.*

Sketch of proof. Let $\Omega = \Lambda^{\mathbf{Z}}$ and let $\mathcal{C}$ be the algebra generated by cylinder sets. For a cylinder set such as $C = \{(x_i)_{i=-\infty}^{\infty} : x_0 = c, x_1 = a, x_2 = t\}$, put $p(C) = P(X_0 = c, X_1 = a, X_2 = t)$.[19]

106. Exercise. Show that p extends uniquely to a premeasure on $\mathcal{C}$. Let $\mathcal{A}$ and μ be the resulting σ-algebra and measure extending p obtained via Carathéodory's theorem. Note that $(\Omega, \mathcal{A}, \mu)$ is a Lebesgue space. $\qquad\square$

107. Aside. According to the previous theorem, if we're dealing with a preprocess $(X_i)_{i=-\infty}^{\infty}$ and whatever we would like to know about it is characterized by the joint distributions of finite collections of the X_i, we can assume it is a process and hence that the space we are working on is Lebesgue. On the other

---

[18] In the notation of the previous footnote, a cylinder set $C(f_1, \ldots, f_r, \lambda_1, \ldots, \lambda_r)$.

[19] More generally, if $C = C(f_1, \ldots, f_r, \lambda_1, \ldots, \lambda_r) = \{(x_i)_{i=-\infty}^{\infty} \in \Omega : x_{f_j} = \lambda_j, 1 \leq j \leq r\}$, define $p(C) = P(X_{f_j} = \lambda_j, 1 \leq j \leq r)$.

hand, beyond their finite joint distributions, preprocesses can be pathological, as we will demonstrate in the remainder of this section. Since, however, none of this material will be used in the rest of the book, disinterested readers may feel free to skip.

108. Example (A pathological preprocess). The first step is to get our hands on a subset of the reals that is not Lebesgue measurable. Here is one way to do this.

109. Exercise. Let $A \subset \mathbf{R}$ be Lebesgue measurable with $m(A) > 0$.

(a) For every $\epsilon > 0$ there exists an interval $I \subset \mathbf{R}$ such that $\frac{\mu(I \cap A)}{|I|} > 1 - \epsilon$.

(b) There exists $\delta > 0$ such that $[-\delta, \delta] \subset A - A = \{x - y : x, y \in A\}$.

110. Exercise. Let G be the additive group of reals generated by $\{1, \pi\}$, let H be the group generated by $\{1, 2\pi\}$, let C be a set of coset representatives for $\frac{\mathbf{R}}{G}$ (note that the existence of such a set requires the axiom of choice) and let $A = C + H = \{c + h : c \in C, h \in H\}$. Show that:

(a) G is dense in $\mathbf{R}$;

(b) $\mathbf{R}$ is the disjoint union of A and $B = A + \pi$;

(c) $A - A$ (and hence $B - B$ as well) does not contain any non-trivial interval centered at 0; hence

(d) for any interval I, $m^*(I \cap A) = m^*(I \cap B) = m(I)$ and $m_*(I \cap A) = m_*(I \cap B) = 0$.

Now that we've got a non-measurable set, we construct a stationary process that acts locally like coin tossing, in the sense that the joint distribution of any finite collection of its constituent random variables models coin tossing efficaciously, but acts globally very unlike coin tossing in that if you interpreted it as such, one of the "tosses" would be wholly determined by the others.

111. Exercise. Justify the following steps:

(a) There exists a set $A \subset [0, 1]$ such that $m^*(A) = 1$ and $m_*(A) = 0$. *Hint: use the foregoing exercise.*

(b) Let $\mathcal{A}$ be the σ-algebra of subsets of $[0, 1]$ generated by $A \cup \mathcal{L}$ (recall that $\mathcal{L}$ is the σ-algebra of Lebesgue measurable subsets of $[0, 1]$). Show that every member of $\mathcal{A}$ can be uniquely written in the form $(A \cap L_1) \cup (A^c \cap L_2)$, where $L_1, L_2 \in \mathcal{L}$.

(c) For $B = (A \cap L_1) \cup (A^c \cap L_2)$, define $\mu(B) = \frac{1}{2}(m(L_1) + m(L_2))$. Then $([0, 1], \mathcal{A}, \mu)$ is a probability space.

(d) The identity map $x \rightarrow x$ is a measure-preserving injection from $([0, 1], \mathcal{A}, \mu)$ to $([0, 1], \mathcal{L}, m)$ that has a non-measurable inverse. Conclude that $([0, 1], \mathcal{A}, \mu)$ is not a Lebesgue space.

(e) A.e. $x \in [0, 1]$ has a unique binary decimal expansion; that is, for a.e. x, there is a unique $\{0, 1\}$-valued sequence $(a_i)_{i=1}^{\infty}$ such that $x = \sum_{i=1}^{\infty} a_i 2^{-i}$. Define (for these values x) a sequence of random variables $(Y_i)_{i=-\infty}^{\infty}$ from $([0, 1], \mathcal{A}, \mu)$ into the alphabet $\Lambda = \{0, 1\}$ as follows:

   (i) $Y_i(x) = a_{2i-1}$ for $i > 0$;
   (ii) $Y_i(x) = a_{-2i}$ for $i < 0$;
   (iii) $Y_0(x) = 1$ if $x \in A$ and 0 otherwise.

   Prove that $(Y_i)_{i=-\infty}^{\infty}$ forms a stationary preprocess.
(f) Let $\mathcal{C} = \{A \cap L : L \in \mathcal{L}\}$ and define a measure ν on $\mathcal{C}$ by $\nu(A \cap L) = m(L)$. Show that the injection map $x \to x$ from $(A, \mathcal{C}, \nu)$ to $([0, 1), \mathcal{L}, m)$ is measure-preserving with a non-measurable inverse. Conclude that $(A, \mathcal{C}, \nu)$ is not a Lebesgue space.
(g) Define a sequence of random variables $(Z_i)_{i=-\infty}^{\infty}$ from $(A, \mathcal{C}, \nu)$ into the alphabet $\Lambda = \{0, 1\}$ as follows: For a.e. $z \in A$, write $z = \sum_{i=1}^{\infty} a_i 2^{-i}$ and put

   (i) $Z_i(x) = a_{2i+1}$ for $i \geq 0$;
   (ii) $Z_i(x) = a_{-2i}$ for $i < 0$.

   Prove that $(Z_i)_{i=-\infty}^{\infty}$ forms a stationary preprocess.
(h) Show that for any $f_1, \ldots, f_r \in \mathbf{Z}$, the joint distributions of $Y_{f_1}, \ldots, Y_{f_r}$ and $Z_{f_1}, \ldots, Z_{f_r}$ are equal to the joint distribution of $X_{f_1}, \ldots, X_{f_r}$, where $(X_i)_{i=-\infty}^{\infty}$ is the coin tossing stationary process defined in Exercise 103 above.
(i) Argue that knowledge of all the X_i, $i \neq 0$, provides no knowledge of the value of X_0, but that knowledge of all the Y_i, $i \neq 0$, determines the value of Y_0 with probability 1.
(j) Argue that all possible sequences of 0s and 1s are equally likely to occur as output of the process $(X_i)_{i=-\infty}^{\infty}$, but that a non-negligible set of sequences is forbidden as output of $(Z_i)_{i=-\infty}^{\infty}$. (What is the meaning of "non-negligible" here?)

112. Discussion. In what sense, exactly, would black boxes housing the two processes $(X_i)_{i=-\infty}^{\infty}$ and $(Y_i)_{i=-\infty}^{\infty}$ be distinguishable one from the other? On one hand, if you knew the non-measurable set A used to construct $(Y_i)_{i=-\infty}^{\infty}$, they would be very distinguishable, for as was mentioned, Y_0 is determined by the other Y_is, this dependence being linked critically to A. On the other hand, if you didn't know (or even know about) A, you probably wouldn't be able to figure out enough about A or catch any inkling of its existence by looking at countably many instances of output from the black box housing $(Y_i)_{i=-\infty}^{\infty}$. Then again, if you could look at an uncountable number of instances of output, you'd probably be able to eventually figure out what A was. (Readers are encouraged to compose for themselves a similar dialectical argument about $(Z_i)_{i=-\infty}^{\infty}$.)

At any rate, it is time to leave pathology behind.

113. Convention. From this point forward, all the probability spaces with which we deal are assumed to be non-atomic Lebesgue spaces unless otherwise noted.

1.5. Conditional expectation

We give basic definitions on conditional expectation. For the purposes of this section, the reader is expected to know the Radon–Nikodym theorem (see, e.g., Folland 1984, Theorem 3.8).

Let $(\Omega, \mathcal{A}, \mu)$ be a Lebesgue space, let f be an integrable function on Ω and suppose that $\mathcal{B} \subset \mathcal{A}$ is a σ-algebra. Form (see Example 56) the canonical factor space $(\Omega', \mathcal{A}', \mu')$ with canonical factor map $\pi : \Omega \to \Omega'$. Now define a measure ν on $(\Omega', \mathcal{A}')$ by the rule $\nu(A') = \int_{\pi^{-1}(A')} f \, d\mu$, $A' \in \mathcal{A}'$.

114. Exercise. Show that ν is non-singular with respect to μ': that is, if $\mu'(A') = 0$ then $\nu(A') = 0$. Conclude by the Radon–Nikodym theorem that there is some μ'-integrable function g such that $\nu(A') = \int_{A'} g \, d\mu'$ for every $A' \in \mathcal{A}'$.

115. Definition. The function $h = g \circ \pi$, where g is as constructed above, is called the *conditional expectation of f given the σ-algebra* $\mathcal{B}$. We denote this function by $h = E(f|\mathcal{B})$.

If $A \in \mathcal{A}$ then the *conditional probability* of A given the σ-algebra $\mathcal{B}$ is just the function $P(A|\mathcal{B})(x) = E(1_A|\mathcal{B})(x)$.

116. Exercise. Show that $E(f|\mathcal{B})$ is $\mathcal{B}$-measurable, is constant on fibers and that for every $B \in \mathcal{B}$, $\int_B E(f|\mathcal{B}) \, d\mu = \int_B f \, d\mu$.

The alert reader will have noticed that since the construction of conditional expectation depended on the construction of Example 56, which in turn depended on the arbitrary choice of a countable dense sequence $(B_i)_{i=1}^{\infty} \subset \mathcal{B}$, there may be an issue about well-definition. Indeed there is, however as the following exercise shows, this issue may be satisfactorily resolved up to sets of measure zero.

117. Exercise. Show that if h_1 and h_2 are $\mathcal{B}$-measurable functions and $\int_B h_1 \, d\mu = \int_B h_2 \, d\mu$ for every $B \in \mathcal{B}$ then $h_1 = h_2$ a.e. Conclude that:

(a) $E(f|\mathcal{B})$ is well defined up to sets of measure zero;
(b) if f is $\mathcal{B}$-measurable then $E(f|\mathcal{B}) = f$ a.e.

118. Definition. A function will be called a *version* of $E(f|\mathcal{B})$ if it coincides a.e. with some (and therefore any) h constructed as above.

119. Example. Let B be some measurable set, let P be a finite partition and let $\mathcal{B}$ be the σ-algebra generated by P. Then if f is an integrable function, $E(f|\mathcal{B})$ takes the constant value $\frac{1}{\mu(p)} \int_p f \, d\mu$ on p for each positive measure $p \in P$. This is of course just the average value of f on p.

 The intuition this example suggests is that $E(f|\mathcal{B})(x)$ is the expected value of $f(x)$, given that you know for every $B \in \mathcal{B}$ whether or not $x \in B$. A further example illustrates this principle in a somewhat different way.

120. Example. Let $(\Omega, \mathcal{A}, \mu)$ be the unit square with Lebesgue measure (of course we didn't actually define this here but we take it that it's more or less standard fare for graduate courses in real analysis) and let $\mathcal{B}$ be the σ-algebra of "vertical sets" $\{B \times [0, 1) : B \in \mathcal{L}\}$. Then for an integrable function f defined on the unit square, a version of $E(f|\mathcal{B})$ is given by $E(f|\mathcal{B})(x, y) = \int f(x, y) \, dm(y)$. It is important to notice that $E(f|\mathcal{B})$ is constant on each vertical fiber $\{x\} \times [0, 1)$.

121. Comment. The reader should take careful note of the general picture implied by the foregoing example when $f = 1_A$ for some measurable A. The whole space $(\Omega, \mathcal{A}, \mu)$ may be represented as a square, with A occupying some region (one does well to imagine a region bounded by a simple closed curve) of two-dimensional space. To get the value of $E(1_A|\mathcal{B})$ at a point (x, y), you just look to see how much of the fiber over (x, y) (namely $\{x\} \times [0, 1)$) is occupied by A. Not all the details of this picture are completely valid in general (the fibers over the points here are all isomorphic copies of $[0, 1)$, which needn't always happen), however there are lots of cases when it is accurate and it's harmless to instruct one's intuition by means of this general image in any case. This image is: represent the whole space $(\Omega, \mathcal{A}, \mu)$ as a square $X \times X$, think of the sub-σ-algebra $\mathcal{B}$ as the algebra of vertical sets $B \times X$, think of the canonical factor space $(\Omega', \mathcal{A}', \mu')$ as the copy of X lying at the base of the horizontal axis and think of the fiber over a point (x, y) as the set $\{\omega\} \times X$. Now to get the value of $E(f|\mathcal{B})$ at a point (x, y), you should think of integrating f against the measure $\mu_{(x,y)}$ on the fiber $\{x\} \times X$. According to a classical theorem of Rohlin (1952), all of this can be formalized. That lies outside the scope of this book but the general picture is instructive nonetheless.

122. Definition. Let X be a measurable map on a Lebesgue space $(\Omega, \mathcal{A}, \mu)$ into a measurable space $(\Lambda, \mathcal{C})$. The *σ-algebra generated by* X is the σ-algebra $\mathcal{B}(X)$ generated by $\{X^{-1}(C) : C \in \mathcal{C}\}$.

Let X, Y, Z or more be measurable maps from a Lebesgue space $(\Omega, \mathcal{A}, \mu)$ into a measurable space $(\Lambda, \mathcal{C})$. The *σ-algebra generated by* X, Y, Z is the σ-algebra $\mathcal{B}(X, Y, Z)$ generated by

$$\{X^{-1}(C) : C \in \mathcal{C}\} \cup \{Y^{-1}(C) : C \in \mathcal{C}\} \cup \{Z^{-1}(C) : C \in \mathcal{C}\}.$$

123. Definition. Let $(\Omega, \mathcal{A}, \mu)$ be a Lebesgue space and let f be an integrable function on Ω. For random variables X, Y, Z or more, defined on Ω, the conditional expectation of f with respect to the σ-algebra $\mathcal{B}(X, Y, Z)$ generated by X, Y and Z is denoted $E(f|X, Y, Z)$.[20]

We'll be needing the following exercises later on.

124. Exercise. Suppose $(f_i)_{i=1}^{\infty}$ is a sequence of functions whose sum converges in a dominated way.[21] Show $E(\sum_{i=1}^{\infty} f_i|\mathcal{B}) = \sum_{i=1}^{\infty} E(f_i|\mathcal{B})$ a.e.

125. Exercise. Show that if $\mathcal{B}_1 \subset \mathcal{B}_2$ are σ-algebras then $E(E(f|\mathcal{B}_1)|\mathcal{B}_2) = E(f|\mathcal{B}_2)$ a.e.

126. Exercise.

(a) Suppose that $f(x) \leq c$ a.e. Prove that for any σ-algebra $\mathcal{B}$, one has $E(f|\mathcal{B})(x) \leq c$ a.e.
(b) Let $\mathcal{B}_2 \subset \mathcal{B}_1$ be σ-algebras and suppose $P(A|\mathcal{B}_1)(x) \leq c$ a.e. Then $P(A|\mathcal{B}_2)(x) \leq c$ a.e. *Hint: for (b), let $f = P(A|\mathcal{B}_1)$ and $\mathcal{B} = \mathcal{B}_2$. Apply part (a) and the previous exercise.*

---

[20] That is, $E(f|X, Y, Z) = E(f|\mathcal{B}(X, Y, Z))$.
[21] In other words, there is some integrable g such that for all n, $|\sum_{i=1}^{n} f_i(x)| < g(x)$ a.e.

2

Measure-preserving systems, stationary processes

2.1. Systems and homomorphisms

In this subchapter, we give basic definitions concerning measure-preserving systems and homomorphisms between them.

127. Definition. Let $(\Omega, \mathcal{A}, \mu)$ be a probability space and assume that $T : \Omega \to \Omega$ is a measure-preserving transformation. We call the quadruple $(\Omega, \mathcal{A}, \mu, T)$ a *measure-preserving system*. If there are sets $X, X' \in \mathcal{A}$ of full measure such that T is a bimeasurable bijection between X and X' then we say that the system $(\Omega, \mathcal{A}, \mu, T)$ is *invertible*, or simply that T is invertible.

128. Comment. Whereas probability theory is the study of probability spaces, ergodic theory is the study of measure-preserving systems. In other words, the most basic object of study for a probabilist is $(\Omega, \mathcal{A}, \mu)$, while the most basic object of study for an ergodic theorist is $(\Omega, \mathcal{A}, \mu, T)$.

129. Convention. In this book we will primarily deal with invertible measure-preserving systems. Accordingly, we may not always say "invertible" though we generally mean it unless we specify otherwise.

130. Definition. Let $(\Omega, \mathcal{A}, \mu, T)$ and $(\Omega', \mathcal{A}', \mu', T')$ be measure-preserving systems and assume that $\pi : \Omega \to \Omega'$ is a measure-preserving transformation such that $T'\pi\omega = \pi T\omega$ for a.e. $\omega \in \Omega$. Then we call π a *homomorphism*. We also say that the system $(\Omega', \mathcal{A}', \mu', T')$ is a *factor* of the system $(\Omega, \mathcal{A}, \mu, T)$, and that the system $(\Omega, \mathcal{A}, \mu, T)$ is an *extension* of the system $(\Omega', \mathcal{A}', \mu', T')$.

131. Definition. Let $(\Omega, \mathcal{A}, \mu, T)$ and $(\Omega', \mathcal{A}', \mu', T')$ be measure-preserving systems and assume that $\pi : \Omega \to \Omega'$ is a homomorphism. If there exist full measure sets $X \subset \Omega$ and $X' \subset \Omega'$ such that the restriction of π to X is a bimeasurable bijection between X and X', we say that π is an *isomorphism* and that the systems $(\Omega, \mathcal{A}, \mu, T)$ and $(\Omega', \mathcal{A}', \mu', T')$ are *isomorphic*.

132. Definition. Let $(\Omega, \mathcal{A}, \mu, T)$ be a measure-preserving system and let $\mathcal{B} \subset \mathcal{A}$ be a sub-σ-algebra that is complete with respect to μ. We say that $\mathcal{B}$ is *T-invariant* if for every $B \in \mathcal{B}$, $T^{-1}B \in \mathcal{B}$.

133. Exercise. Show that if $\pi : \Omega \to \Omega'$ is a homomorphism then $\{\pi^{-1}(A) : A \in \mathcal{A}'\}$ is a T-invariant σ-algebra of subsets of Ω.

134. Exercise. Let $(\Omega, \mathcal{A}, \mu, T)$ be a measure-preserving system on a Lebesgue space and suppose that $\mathcal{B} \subset \mathcal{A}$ is a T-invariant σ-algebra. Carry out the construction of the canonical factor space $(\Omega', \mathcal{A}', \mu')$.[22] Show that T projects to a measure-preserving transformation T' on Ω' and that $(\Omega', \mathcal{A}', \mu', T')$ is a factor of $(\Omega, \mathcal{A}, \mu, T)$.

135. Convention. When $(\Omega, \mathcal{A}, \mu, T)$ is a measure-preserving system and $\mathcal{B} \subset \mathcal{A}$ is a T-invariant σ-algebra we shall refer to $\mathcal{B}$ as a factor; what we have in mind is the system $(\Omega', \mathcal{A}', \mu', T)$ of the foregoing exercise.

136. Definition. A system $(\Omega, \mathcal{A}, \mu, T)$ is said to be *ergodic* if there is no measurable set A of measure strictly between 0 and 1 such that $\mu(A \triangle T^{-1}A) = 0$.

137. Exercise. Show that $(\Omega, \mathcal{A}, \mu, T)$ is ergodic if and only if $A \in \mathcal{A}$ and $A = T^{-1}A$ implies $\mu(A) \in \{0, 1\}$.

2.2. Constructing measure-preserving transformations

In this subchapter we outline three basic ways to construct measure-preserving transformations. For the purposes of this book, only the second and third are essential.

138. First method: explicitly defined functions. Take Ω to be some set, say for example $[0, 1)$, and define some function $T : \Omega \to \Omega$ by an explicit formula, for example, $Tx = e^x$ mod 1. This T doesn't preserve Lebesgue measure, but it does preserve some measure, for example the point measure[23] on any point y that is a solution to the equation $x = e^x$ mod 1. (We leave it to the reader to prove that there is such a solution to this equation in $[0, 1)$.) This suggests a general scheme whereby one constructs a measure-preserving system by first specifying a set Ω and a self-map T and then finding a measure preserved by T. That this will work under fairly general circumstances is the content of the following exercise.

---

[22] See Example 56 and the series of exercises following it.

[23] The point measure on a point y is the measure that assigns 1 to any set containing y and 0 to all other sets.

139. Exercise. Let X be a compact metric space and suppose $T : X \to X$ is continuous. Prove there exists a measure μ on the Borel σ-algebra $\mathcal{A}$ such that $(\Omega, \mathcal{A}, \mu, T)$ is a measure-preserving system by completing the steps in the following argument. (Note: this exercise requires some functional analytic background. The reader is encouraged to look to the relevant sources for the details, but we will not use the contents of this exercise in the rest of the book.)

(a) The space $M(X)$ of probability measures on $\mathcal{A}$ can be identified with $\{\lambda \in C(X)^* : C(1) = 1, C(f) \geq 0 \text{ if } f \geq 0\}$. Namely, μ is identified with λ when $\int f \, d\mu = \lambda(\mu)$. (This is the Riesz representation theorem; its proof is non-trivial. See for example Folland 1984, Theorem 7.17.)

(b) $M(X)$ is compact in the weak* topology. (This follows from the fact that the unit ball of $C(X)^*$ is compact and $M(X)$ is closed in the weak* topology; see Folland 1984, Theorems 5.18, 7.17.)

(c) If $\mu \in M(X)$, let $T\mu$ be the measure defined by $\int f \, dT\mu = \int Tf \, d\mu$. Now fix any $\sigma \in M(X)$ and let μ be any weak* limit point of the sequence $\frac{1}{n} \sum_{i=1}^{n} T^n \sigma$. Show that $\mu \in M(X)$ is T-invariant.

140. Second method: cutting and stacking. This is a very informal and intuitive introduction to cutting and stacking constructions. We'll use almost no notation. Start with an interval taken from the real line and cut the interval into a bunch of sub-intervals of equal length. Stack these sub-intervals vertically and define a transformation on the whole stack except for the top "rung" to be the map that sends any point to the point immediately above it (on the next level up).

141. Example. Say the interval you start with is $\left[0, \frac{3}{4}\right)$, which you cut into three pieces $\left[0, \frac{1}{4}\right)$, $\left[\frac{1}{4}, \frac{2}{4}\right)$ and $\left[\frac{2}{4}, \frac{3}{4}\right)$. After stacking them vertically, the picture looks something like this:

$$\left[\frac{2}{4}, \frac{3}{4}\right)$$
$$\left[\frac{1}{4}, \frac{2}{4}\right)$$
$$\left[\frac{0}{4}, \frac{1}{4}\right)$$

The transformation at this stage takes x to $x + \frac{1}{4}$ for $x \in \left[0, \frac{2}{4}\right)$ and is undefined for $x \in \left[\frac{2}{4}, \frac{3}{4}\right)$, that is, on the top rung. To define it on (at least part of) the top rung, cut the whole tower vertically into columns of equal width and stack them into one or more taller, narrower towers. While you're at it, you can add extra rungs (called *spacers*) in between these columns. These spacers are new intervals from the real line, disjoint from each other and any previously chosen; adding them increases the measure of the space you're constructing.

The transformation still sends a point on any rung (not the top rung, though) to the point directly above it on the next rung up. You should check that this means it's defined the same way it was before, where it was defined before.

142. Example. Take the previous stack and split it into five equal sized columns, like this:

$[\frac{10}{4}, \frac{11}{4})$	$[\frac{11}{4}, \frac{12}{4})$	$[\frac{12}{4}, \frac{13}{4})$	$[\frac{13}{4}, \frac{14}{4})$	$[\frac{14}{4}, \frac{15}{4})$
$[\frac{5}{20}, \frac{6}{20})$	$[\frac{6}{20}, \frac{7}{20})$	$[\frac{7}{20}, \frac{8}{20})$	$[\frac{8}{20}, \frac{9}{20})$	$[\frac{9}{20}, \frac{10}{20})$
$[\frac{0}{20}, \frac{1}{20})$	$[\frac{1}{20}, \frac{2}{20})$	$[\frac{2}{20}, \frac{3}{20})$	$[\frac{3}{20}, \frac{4}{20})$	$[\frac{4}{20}, \frac{5}{20})$

Now arrange these columns into two stacks as follows. The first stack consists of (in this order) the first column, three spacers, then the second column. The second column consists of the third column, the fourth column, a spacer and the fifth column. We'll give two pictures of the situation now, one with the spacers labeled as such and a second one in which the spacers are labeled as new intervals from the real line.

Left picture:

	$[\frac{14}{4}, \frac{15}{4})$
$[\frac{11}{4}, \frac{12}{4})$	$[\frac{9}{20}, \frac{10}{20})$
$[\frac{6}{20}, \frac{7}{20})$	$[\frac{4}{20}, \frac{5}{20})$
$[\frac{1}{20}, \frac{2}{20})$	spacer
spacer	$[\frac{13}{4}, \frac{14}{4})$
spacer	$[\frac{8}{20}, \frac{9}{20})$
spacer	$[\frac{3}{20}, \frac{4}{20})$
$[\frac{10}{4}, \frac{11}{4})$	$[\frac{12}{4}, \frac{13}{4})$
$[\frac{5}{20}, \frac{6}{20})$	$[\frac{7}{20}, \frac{8}{20})$
$[\frac{0}{20}, \frac{1}{20})$	$[\frac{2}{20}, \frac{3}{20})$

or

Right picture:

	$[\frac{14}{4}, \frac{15}{4})$
$[\frac{11}{4}, \frac{12}{4})$	$[\frac{9}{20}, \frac{10}{20})$
$[\frac{6}{20}, \frac{7}{20})$	$[\frac{4}{20}, \frac{5}{20})$
$[\frac{1}{20}, \frac{2}{20})$	$[\frac{18}{4}, \frac{19}{4})$
$[\frac{17}{4}, \frac{18}{4})$	$[\frac{13}{4}, \frac{14}{4})$
$[\frac{16}{4}, \frac{17}{4})$	$[\frac{8}{20}, \frac{9}{20})$
$[\frac{15}{4}, \frac{16}{4})$	$[\frac{3}{20}, \frac{4}{20})$
$[\frac{10}{4}, \frac{11}{4})$	$[\frac{12}{4}, \frac{13}{4})$
$[\frac{5}{20}, \frac{6}{20})$	$[\frac{7}{20}, \frac{8}{20})$
$[\frac{0}{20}, \frac{1}{20})$	$[\frac{2}{20}, \frac{3}{20})$

Continue this process of cutting and stacking countably many times, making sure that your transformation ends up being defined almost everywhere. (In other words, so that the combined measure of the tops of the stacks converges to zero.) Also you want to make sure that the whole space is finite in measure, so you can't add arbitrarily many spacers at each stage. At the end, normalize so that the measure of the whole space is 1.

143. Third method: stationary processes. A stationary processes $(X_i)_{i=-\infty}^{\infty}$ on an alphabet Λ gives rise to a measure-preserving transformation in the following way. Recall that $\Omega = \Lambda^{\mathbf{Z}}$ is the domain space of the process. Define a map $T : \Omega \to \Omega$ by $T\big((x_i)_{i=-\infty}^{\infty}\big) = (z_i)_{i=-\infty}^{\infty}$, where $z_i = x_{i+1}$.

144. Exercise. Show that $(\Omega, \mathcal{A}, \mu, T)$ is an invertible measure-preserving system. *Hint: first show that the collection of cylinder sets generates $\mathcal{A}$ mod 0.*

In this book, most of the concrete measure-preserving transformations we construct will be obtained by way of the third method, i.e. via stationary processes.

2.3. Types of processes; ergodic, independent and (P, T)

The material in this subchapter is essential.

145. Exercise. Let $(X_i)_{i=-\infty}^{\infty}$ and $(Y_i)_{i=-\infty}^{\infty}$ be stationary processes. If for any $r \in \mathbf{N}$ and $f_1, \ldots, f_r \in \mathbf{Z}$, the joint distribution of $X_{f_1}, \ldots, X_{f_r}$ is equal to the joint distribution of $Y_{f_1}, \ldots, Y_{f_r}$, then the processes $(X_i)_{i=-\infty}^{\infty}$ and $(Y_i)_{i=-\infty}^{\infty}$ are isomorphic.

146. Comment. We may at times talk about stationary process being ergodic, isomorphic to measure-preserving systems or other processes, etc. In these cases, one can infer the meaning by taking the assertion to be about the associated measure-preserving system.

A typical example of a stationary process that fails to be ergodic arises as follows. Take two coins, one a fair $\frac{1}{2}$-$\frac{1}{2}$ coin and the other one an unfair coin, say $\frac{2}{3}$-$\frac{1}{3}$. Randomly pick one of them, each with probability $\frac{1}{2}$, and use it to produce a doubly infinite sequence of heads and tails. This gives rise to a stationary process by analogy with Exercise 106.[24] This stationary process gives rise, in turn, to a measure-preserving system $(\Omega, \mathcal{A}, \mu, T)$. This system fails to be ergodic, for example because (see the Birkhoff ergodic theorem below) the set $A = \Big\{ (\omega_i)_{i=-\infty}^{\infty} \in \Omega : \lim_{n \to \infty} \frac{|\{i \in [-n,n] : \omega_i = h\}|}{2n+1} = \frac{1}{2} \Big\}$ satisfies $\mu(A) = \frac{1}{2}$ and $\mu(A \triangle T^{-1} A) = 0$.

From the perspective of pure mathematics, non-ergodic processes are somewhat unnatural, and we won't treat them here. Note, however, that they do arise in some applications; for such cases, ergodic decomposition (see below) can be a useful tool.

147. Comment. Notice that the definition of "ergodic" refers to $T^{-1} A$ rather than $T A$. This distinction is substantive when T is not invertible, as in this case, $T A$ may often not have the same measure as A. (Consider for example the map

---

[24] You just have to figure out the relative frequencies of all finite words, in order to construct the required premeasure. For starters, you can verify that the relative frequencies of hh, ht, th and tt are $\frac{25}{72}$, $\frac{17}{72}$, $\frac{17}{72}$ and $\frac{13}{72}$, respectively.

$Tx = 2x$ (mod 1), which preserves Lebesgue measure on $[0, 1)$.) Even when T is invertible, however, we will frequently prefer to write, e.g. $T^{-n}A$ rather than $T^n A$. This is because $T^n \omega \in A$ if and only if $\omega \in T^{-n}A$.

148. Definition. A stationary process $(X_i)_{i=-\infty}^{\infty}$ is called an *independent process* if the random variables X_i are independent of each other, so that for example $P(X_0 = a, X_1 = b, X_2 = c) = P(X_0 = a)P(X_0 = b)P(X_0 = c)$.

149. Theorem. *Any independent process is ergodic.*

Sketch of Proof. Let $(\Omega, \mathcal{A}, \mu, T)$ be the system arising from the process. Suppose that this system is not ergodic. Then there exists $A \in \mathcal{A}$ with $0 < \mu(A) < 1$ such that $\mu(A \triangle T^{-1}A) = 0$. By Corollary 71, the algebra of cylinder sets generates $\mathcal{A}$ mod 0.

150. Exercise. For any $\epsilon > 0$ there exist two cylinder sets C_1 and C_2 such that each C_i approximates A up to ϵ and yet C_1 and C_2 are independent. Use this fact to complete the proof. □

Formerly we saw how a process gives rise to a measure-preserving transformation. Now we shall see how to run this correspondence in reverse.

151. Definition. Let $(\Omega, \mathcal{A}, \mu, T)$ be a measure-preserving system and let P be a countable, measurable partition. The (P, T) *process* is the stationary process $\ldots X_{-3}, X_{-2}, X_{-1}, X_0, X_1, X_2, X_3, \ldots$ whose alphabet consists of the pieces of P, such that X_i takes a point ω to the value p, if p is the member of P containing $T^i \omega$.[25]

152. Discussion. Given a stationary process $(X_i)_{i=-\infty}^{\infty}$, where the X_i take values in a countable alphabet $\Lambda = \{\lambda_1, \lambda_2, \ldots\}$, one can form the associated measure-preserving shift system $(\Omega, \mathcal{A}, \mu, T)$ per Exercise 99, then let P be the partition consisting of the pieces $\{x : x_0 = \lambda_i\}$, $i = 1, 2, \ldots$, and construct the (P, T) process, which will be isomorphic to the original process and, for all practical purposes, indistinguishable from it. That is why we can always assume that the underlying space for our stationary processes $(X_i)_{i=-\infty}^{\infty}$ is $\Omega = \Lambda^{\mathbf{Z}}$ and that the underlying transformation is the shift. Notice that, modulo this assumption, when the alphabet Λ is fixed, *a stationary process is completely characterized by the measure* μ. This is extremely important, and the reader should make careful note of it.

153. Exercise. Show that the (P, T) process is stationary.

---

[25] To be more precise: let $P = \{P_1, P_2, \ldots\}$ be a countable, measurable partition of Ω. Choose an alphabet $\Lambda = \{\lambda_1, \lambda_2, \ldots\}$ having the same cardinality as P and for $\omega \in \Omega$ and $i \in \mathbf{Z}$, define $X_i(\omega) = \alpha_j$ if $T^i \omega \in P_j$. $(X_i)_{i=-\infty}^{\infty}$ is the (P, T) process.

154. Definition. If the (P, T) process separates points mod 0,[26] we say that P *generates* T, or that P is a *generator* for T.

155. Exercise. Show that P generates T if and only if $\{T^i p : p \in P, i \in \mathbf{Z}\}$ generates $\mathcal{A}$ mod 0.

156. Definition. Let $(\Omega, \mathcal{A}, \mu, T)$ be a measure-preserving system, let P be a countable partition of Ω and let $\omega \in \Omega$. The *P-name* of ω is the sequence $(p_i)_{i=-\infty}^{\infty}$, where for each i, p_i is the member of P containing $T^i \omega$. If $n \in \mathbf{N}$, the *P-name of length* n (or simply the *n-name*, when P is understood) of ω is the finite sequence $(p_i)_{i=0}^{n-1}$.

157. Comment. In practice, one usually chooses an alphabet $\Lambda = \{\lambda_1, \lambda_2, \ldots\}$ in one-to-one correspondence with a partition $P = \{p_1, p_2, \ldots\}$ and writes the P-name of ω as $(\lambda_i)_{i=-\infty}^{\infty}$.

158. Theorem. *If P generates T then the (P, T) process is isomorphic[27] to $(\Omega, \mathcal{A}, \mu, T)$.*

Sketch of proof. Let π be the map taking ω to the P-name of ω.[28]

159. Exercise. Show that π is an isomorphism.   □

160. Corollary. *If two partitions P and Q each generate T then the (P, T) process is isomorphic to the (Q, T) process.*

2.4. Rohlin tower theorem

In this section we give three versions of the Rohlin tower theorem, which is one of the fundamental tools of constructive ergodic theory.

161. Definition. Let $(\Omega, \mathcal{A}, \mu)$ be a probability space. Write $A \sim B$ if $\mu(A \triangle B) = 0$.

162. Exercise. Show that $\sim$ is an equivalence relation.

163. Definition. For $A \in \mathcal{A}$, write $\overline{A}$ for the equivalence class of A under $\sim$ and write $\overline{\mathcal{A}}$ for the family of equivalence classes.

164. Definition. For $\overline{A}, \overline{B} \in \overline{\mathcal{A}}$, write $\overline{A} \preceq \overline{B}$ if $\mu(A \setminus B) = 0$.

---

[26] This means that there exists a null set E such that for every pair of distinct points $x, y \in \Omega \setminus E$ there is some $i \in \mathbf{Z}$ such that $T^i x$ and $T^i y$ lie in different cells of P.

[27] That is to say, the measure-preserving system generated by the (P, T) process.

[28] To be more precise, let $(X_i)_{i=-\infty}^{\infty}$ be the (P, T) process and let $(X, \mathcal{B}, \nu, S)$ be the system arising from this process. (So $X = \Lambda^{\mathbf{Z}}$, where Λ is an alphabet indexing P, S is the shift, etc.) For $\omega \in \Omega$, let $\pi(\omega) = (x_i)_{i=-\infty}^{\infty}$, where $x_i = \lambda$ if and only if $T^i \omega \in p_\lambda$.

165. Exercise. Show that the relation $\preceq$ is well defined (that is, does not depend on the representatives of the classes), and that $\preceq$ is a partial order.[29]

166. Theorem. *(Measurable Zorn lemma.) Let $(\Omega, \mathcal{A}, \mu)$ be a probability space and let $\mathcal{S} \subset \mathcal{A}$ be a collection of sets. If every totally ordered (under $\preceq$) subcollection of $\mathcal{S}$ has an upper bound, then $\mathcal{S}$ has a maximal element.*

167. Convention. We are writing $A \preceq B$ as shorthand for $\overline{A} \preceq \overline{B}$. Note, however, that $\preceq$ is *not* a partial order on $\mathcal{A}$ (it fails antisymmetry). There is a corresponding point about the notion of a maximal element: $C \in \mathcal{S}$ is maximal with respect to $\preceq$ if $D \in \mathcal{S}$ and $C \preceq D$ implies that $D \sim C$ (instead of $D = C$).

Sketch of proof. Pick a chain (S_n) where $\mu(S_n) - \mu(S_{n-1})$ is always at least half of what it could be.[30]

168. Exercise. Let S be an upper bound of (S_n). Show that S is a maximal element. $\qquad\square$

169. Theorem. *(Rohlin tower theorem: ergodic version.) Let $(\Omega, \mathcal{A}, \mu, T)$ be an ergodic measure-preserving system, let $N \in \mathbf{N}$ and let $\epsilon > 0$. There exists some $S \in \mathcal{A}$ such that $S, TS, T^2S, \ldots, T^{N-1}S$ are pairwise disjoint and $\mu(X \setminus \bigcup_{i=0}^{N-1} T^i S) < \epsilon$.*

Idea of proof. *Start with a tiny set C, and let S be the set of all points $T^n x$ such that $x \in C$, n is a non-negative multiple of N, and $\{Tx, T^2x, \ldots, T^{n+N}x\}$ is entirely outside of C. Then $\{S, TS, T^2S, \ldots, T^N S\}$ is a disjoint cover of all of the space except $C \cup T^{-1}C \cup T^{-2}C \cup \cdots \cup T^{-N}C$, because by ergodicity, the translates of C cover the whole space.*

Sketch of proof. Let C be a set of measure less than $\frac{\epsilon}{N}$ and let

$$S = \left\{ T^{kN}\omega : k \in \mathbf{Z}, k \geq 0, \omega \in C, \ C \cap \{T\omega, T^2\omega, T^3\omega, \ldots, T^{(k+1)N}\omega\} = \emptyset \right\}.$$

170. Exercise. Show that $(X \setminus \bigcup_{i=0}^{N-1} T^i S) \subset \bigcup_{i=0}^{N-1} T^{-i}C$. *Hint: by ergodicity, the union of the translates of C covers Ω mod 0.* $\qquad\square$

171. Definition. Let $(\Omega, \mathcal{A}, \mu, T)$ be a measure-preserving system. We say that T is *non-periodic*, or $(\Omega, \mathcal{A}, \mu, T)$ is non-periodic, if for every $i \in \mathbf{N}$ the probability that $T^i x = x$ is zero.[31]

---

[29] That is, a reflexive, antisymmetric and transitive relation.

[30] In other words, for $S \in \mathcal{S}$, let $f(S) = \sup\{\mu(B) : B \in \mathcal{S} \text{ with } S \preceq B\}$. Let $S_1 \in \mathcal{S}$ and for $n > 1$, choose S_n with $S_{n-1} \preceq S_n$ and $\mu(S_n) \geq [\mu(S_{n-1}) + f(S_{n-1})]/2$.

[31] That is, $\mu(\{x : \exists i \in \mathbf{N} \text{ with } T^i x = x\}) = 0$.

172. Theorem. *(Rohlin tower theorem: non-periodic version.) Let* $(\Omega, \mathcal{A}, \mu, T)$ *be a non-periodic measure-preserving system, let* $N \in \mathbf{N}$ *and let* $\epsilon > 0$. *There exists some* $S \in \mathcal{A}$ *such that* $S, TS, T^2S, \ldots, T^{N-1}S$ *are pairwise disjoint and* $\mu(X \setminus \bigcup_{i=0}^{N-1} T^i S) < \epsilon$.

Idea of proof. Since the space is Lebesgue, we can assume it is the unit interval, endowed with its usual metric. Let $M \gg N$. Use the same proof as above, letting (by measurable Zorn) C be a maximal set such that for every $x \in C$, $Tx, T^2x, \ldots, T^Mx$ are all outside C. The only problem is to show that C and its translates cover the whole space. If not, note that by non-periodicity, for δ small enough there is a positive probability that $Tx, T^2x, \ldots, T^Mx$ are all more than δ away from x, and x is not in the union of C and its translates. There must be some interval of diameter δ which intersects that event on a set of positive measure, and that intersection is a set of positive measure outside the union of the translates of C, such that for any x in that set, $Tx, T^2x, \ldots, T^Mx$ are all outside of that set. Add that set to C to contradict maximality.

Sketch of proof.

173. Exercise. Show that it is sufficient (by simply mimicking the proof of the ergodic case) to show that there is a set $C \in \mathcal{A}$ with $\mu(C) < \epsilon/N$ such that the union of the translates of C covers Ω mod 0.   •

We now turn our attention to the construction of the set C.

174. Exercise. Show that we may without loss of generality assume that $(\Omega, \mathcal{A}, \mu)$ is the unit interval with Lebesgue measure. (Recall our convention that $(\Omega, \mathcal{A}, \mu)$ is a Lebesgue space. One needs to rule out atoms.)   •

An advantage of working in the unit interval is that we can exploit its usual metric.

Let M be very large and let S be the family consisting of measurable sets A having the property that $A, T^{-1}A, T^{-2}A, \ldots, T^{-M}A$ are pairwise disjoint.

175. Exercise. Show that every totally ordered (under $\preceq$) subfamily of S has an upper bound (in S).   •

By the foregoing exercise and the measurable Zorn lemma, choose a maximal element C in S. Suppose, for proof by contradiction, that C and all its translates don't cover Ω mod 0.

176. Exercise. There exists $\delta > 0$ such that, with positive probability, x is not in the union of the translates of C and $Tx, T^2x, \ldots, T^M x$ are all more than δ away from x. Call this event E.[32] *Hint: let $F_n = \{x : |x - T^j x| > \frac{1}{n}, 1 \le j \le M\}$ and show that $\mu(F_n) \to 1$.* •

Choose an interval I of diameter $< \delta$ such that $\mu(I \cap E) > 0$. Let $C' = C \cup (I \cap E)$.

177. Exercise. Show that $C' \in S$, $C \preceq C'$ and $C' \not\succeq C$, a contradiction. □

178. Definition. Let $(\Omega, \mathcal{A}, \mu, T)$ be an invertible measure-preserving system and suppose $S \in \mathcal{A}$ with $S, TS, T^2S, \ldots, T^{N-1}S$ pairwise disjoint. We call $\{S, TS, T^2S, \ldots, T^{N-1}S\}$ a *Rohlin tower of height N*. The sets $T^i S$ are called the *rungs* of the tower, and S is called the *base*. The set $\Omega \setminus \bigcup_{i=0}^{N-1} T^i S$ is called the *error set*.

179. Comment. In all arguments, we shall assume that the error set is sufficiently small in measure.

180. Exercise. Prove that, in Theorem 172, one may choose the error set to have measure exactly ϵ.

181. Definition. Let $(\Omega, \mathcal{A}, \mu, T)$ be a measure-preserving system, let P be a countable measurable partition of Ω, let $N \in \mathbf{N}$ and let $\{S, TS, \ldots, T^{N-1}S\}$ be a Rohlin tower. For $x, y \in E$, put $x \sim y$ if x and y have the same P-name of length N. If E is an equivalence class of this relation, the family $\{E, TE, \ldots, T^{N-1}E\}$ is called a *(P, T)-column* of the tower. The sets $T^i E$ are called the *rungs* of the column.

182. Comment. Each rung of a (P, T) column lies entirely inside a member of P.

183. Definition. Let $(\Omega, \mathcal{A}, \mu)$ be a probability space. An event S and a measurable partition $P = \{p_1, p_2, \ldots\}$ are *independent* if S and p_i are independent for each i. If $Q = \{q_1, q_2, \ldots\}$ is another partition, we say that Q and P are independent if p_i and q_j are independent for all i, j.

184. Theorem. *(Rohlin tower theorem: independent base version.) Let $(\Omega, \mathcal{A}, \mu, T)$ be a non-periodic measure-preserving system, let $N \in \mathbf{N}$ and let $\epsilon > 0$. For any finite measurable partition P, there exists some $S \in \mathcal{A}$ that*

---

[32] That is, $E = \{x \in \Omega \setminus \bigcup_{i=-\infty}^{\infty} T^i C : |x - T^j x| > \delta, 1 \le j \le M\}$ and one has $\mu(E) > 0$.

is independent of P such that $S, TS, T^2S, \ldots, T^{N-1}S$ are pairwise disjoint and $\mu(X \setminus \bigcup_{i=0}^{N-1} T^i S) = \epsilon$.

Sketch of proof.[33] Let δ be extremely small and let M be extremely large. Choose a Rohlin tower of height M with error set of measure less than δ. Break each (P, T)-column into N subcolumns. We picture the situation with $N = 4$ below. In the picture you see a single (P, T)-column, split into four vertical subcolumns. You put the pieces with the $\sim$ marks into the set S. Do this for all the (P, T)-columns.

---

[33] Readers are encouraged to provide their own arguments from the intuitive sketch provided; here we offer a few more details. First, note that the independence condition amounts to the requirement that $\mu(S \cap p) = \frac{1-\epsilon}{N}\mu(p)$ for every $p \in P$, and that if one can satisfy $\mu(S \cap p) > \frac{1-\epsilon}{N}\mu(p)$ for every $p \in P$, one can finish the proof by shaving off suitably sized portions of each $S \cap p$. Choose $\delta > 0$ such that $\delta < \mu(p)\epsilon$ for every $p \in P$. Next choose $M > \frac{2N}{\delta}$ and let $B, TB, \ldots, T^{M-1}B$ be a Rohlin tower with an error set F of measure less than $\frac{\delta}{2}$. Now for each (P, T)-column $\{E, TE, \ldots, T^{M-1}E\}$ of this tower, do the following. Let $E = c_0 \cup c_1 \cup \cdots \cup c_{N-1}$ be a partition of E into equal measure pieces and let $S_E = \bigcup_{i=0}^{M-N} T^i c_{m(i)}$, where $m(i)$ is the remainder when i is divided by N. Finally put $S = \bigcup S_E$, where E runs over the bases of the (P, T)-columns in the tower. S is the base of a Rohlin tower of height N; call its error set F'. One must show that $\mu(F') < \delta$. By construction $\mu(S \cap p) = \frac{\mu(p \setminus F)}{N} > \frac{\mu(p)-\delta}{N} > \frac{\mu(p)-\epsilon\mu(p)}{N} = \frac{1-\epsilon}{N}\mu(p)$ for each $p \in P$.

185. Exercise. Show that $\{S, TS, T^2S, \ldots, T^{N-1}S\}$ is a Rohlin tower of height N, and that, if δ is small enough and M large enough, one may, by shaving off a small part of S and throwing it into the error set, achieve independence from P while keeping the error set under ϵ in measure. □

186. Definition. The *superimposition* of a sequence of partitions (P_i) is the partition of Ω into the equivalence classes of $\sim$, where $x \sim y$ if for every $i \in \mathbf{N}$, x and y are in the same cell of P_i.

187. Notation. The superimposition of a sequence of partitions (P_i) will be denoted by $\bigvee_i P_i$. The superimposition of two partitions A and B will be denoted by $A \vee B$.

188. Exercise. Show that in the Rohlin tower theorem you can in fact get the base of the Rohlin tower to be independent of finitely many finite measurable partitions. *Hint: consider the superimposition of the partitions.*

189. Exercise. Use the previous exercise to show that one can in fact get every rung of the tower to be independent of a given finite measurable partition P. *Hint: if the height is to be N, get the base S to be independent of $P, T^{-1}P, T^{-2}P, \ldots, T^{-N}P$.*

190. Comment. Given a (P, T) process, where P is a finite partition, if you pick a Rohlin tower of height n whose base is independent of the (finite) partition of the space by equivalence of n names, you get the beautiful situation where the distribution of names defined by the columns is precisely the distribution of all n names.

2.5. Countable generator theorem

In this subchapter we show that an arbitrary invertible measure-preserving transformation on a Lebesgue space is isomorphic to a stationary process on a countable alphabet.

191. Definition. A measurable partition P is said to be *countable mod 0* if P has a countable subfamily P' such that $\mu(\Omega \setminus \bigcup_{p \in P'}) = 0$.

Suppose now that (P_i) is a sequence of measurable partitions of Ω. For $x, y \in \Omega$ write $x \sim y$ if for every $i \in \mathbf{N}$, x and y are in the same cell of P_i.

192. Exercise. Show that $\sim$ is an equivalence relation and that its equivalence classes form a measurable partition of Ω. (Measurability is primarily of interest when there are equivalence classes of positive measure.)

193. Theorem. *Let $(\Omega, \mathcal{A}, \mu)$ be a probability space and suppose that $(S_i)_{i=1}^{\infty}$ is a sequence of measurable sets with $\sum_{i=1}^{\infty} \mu(S_i) < \infty$. Suppose that for each $i \in \mathbf{N}$, P_i is a finite measurable partition of Ω having the property that $S_i^c \in P_i$. Then the superimposition P of $(P_i)_{i=1}^{\infty}$ is countable mod 0.*

Sketch of proof.

194. Exercise. Show that it is sufficient to show that for each $\epsilon > 0$, there is a finite subfamily P' of P such that $\mu(\bigcup_{p \in P'} p) > 1 - \epsilon$.   ●

Let $\epsilon > 0$ and choose j such that $\mu(\bigcup_{i=j}^{\infty} S_i) < \epsilon$.

195. Exercise. There is a finite subfamily P' of P such that $\left(\Omega \setminus \bigcup_{i=j}^{\infty} S_i\right) \subset \bigcup_{p \in P'} p$.   □

The reasoning of the foregoing proof recalls a famous lemma. Since we'll be needing this lemma anyway, now's a good time to introduce it. The reader is encouraged to look for its covert use in the above proof.

196. Theorem. *(Borel–Cantelli lemma; see e.g. Folland 1984, Lemma 9.10.) Let $(\Omega, \mathcal{A}, \mu)$ be a probability space and suppose $(A_i)_{i=1}^{\infty}$ is a sequence of events satisfying $\sum_{i=1}^{\infty} \mu(A_i) < \infty$. Then for a.e. $\omega \in \Omega$, $\omega \in A_i$ for at most finitely many i.*

197. Exercise. Prove the Borel–Cantelli lemma.   □

198. Exercise. It's possible (indeed typical) for the superimposition of a countable sequence of finite partitions to fail countability mod 0. *Hint: let P_i be partition of $[0, 1)$ into two cells according to the value of the ith binary digit.*

199. Theorem. *(Rohlin 1965.) Every invertible measure-preserving system on a Lebesgue space is isomorphic to a stationary process on a countable alphabet.*

Idea of proof. Again we assume the space to be $[0, 1]$ with Lebesgue measure. All we need is a countable partition of the space so that the map which takes points to the name of points with respect to that partition is an isomorphism, i.e. such that the point-to-name map is one-to-one. To do this, create Rohlin towers with bases $S_1, S_2, \ldots$ so that the measures of the S_i are summable, and with insignificant error sets. Assume the heights of these towers are $N_1, N_2, \ldots$ Let P_i be a partition of the whole space, consisting of the complement of S_i together with the following partition of S_i: two points x and y are in the same atom if and only if for all $k \in \{1, 2, \ldots, N_i\}$, $T^k x$ and $T^k y$ have the same first

N_i digits in their binary expansions. The superimposition of the P_i supplies the desired partition; error sets can be obnoxious but Borel–Cantelli can dispose of all but finitely many of them.

Sketch of proof. Let $(\Omega, \mathcal{A}, \mu, T)$ be the system in question.

200. Exercise. Show that we may, without loss of generality, assume that $(\Omega, \mathcal{A}, \mu)$ is $[0, 1]$ with Lebesgue measure. •

By Theorem 158, it will suffice to find a partition P that is countable mod 0 and that generates T, i.e. (see Definition 154) such that $\{T^i p : i \in \mathbf{Z}, \ p \in P\}$ separates points mod 0.

For each $n \in \mathbf{N}$, let S_n be the base of a Rohlin tower of height n^2 having error set at most $\frac{1}{n^2}$ in measure. For $x, y \in S_n$, write $x \sim y$ if $T^k x$ and $T^k y$ agree in the first n digits of their binary expansions, $0 \le k \le n$. It is easy to see that $\sim$ is an equivalence relation on S_n and that its equivalence classes are measurable. Let P_n be the partition consisting of S_n^c and the equivalence classes of $\sim$. Let P be the superimposition of $(P_i)_{i=1}^\infty$. Since $\sum_{i=1}^\infty \mu(S_i) < \infty$, by Theorem 193, P is countable mod 0. Let E be the intersection of the error sets of the Rohlin towers. Clearly E is a null set.

201. Exercise. Show that for $x, y \in E^c$, there is some $p \in P$ and some $n \in \mathbf{N}$ such that $T^n x \in p$ and $T^n y \notin p$. □

2.6. Birkhoff ergodic theorem and the strong law

We give two versions of the Birkhoff ergodic theorem, possibly the most basic theorem in ergodic theory, and one of its probabilistic variants, the strong law of large numbers.

202. Theorem. *(Birkhoff ergodic theorem; Birkhoff 1931.) Let $(\Omega, \mathcal{A}, \mu, T)$ be a measure-preserving system and suppose $f : \Omega \to \mathbf{R}$ is an integrable function. Then for a.e. $\omega \in \Omega$, $\lim_{N\to\infty} \frac{1}{N} \sum_{n=1}^N f(T^i \omega)$ exists.*

Idea of proof. Fix $b > a$. Consider the set of points ω with

$$\limsup \frac{1}{n}\big(f(\omega) + f(T\omega) + \ldots + f(T^n\omega)\big) > b$$

and

$$\liminf \frac{1}{n}\big(f(\omega) + f(T\omega) + \ldots + f(T^n\omega)\big) < a.$$

Let B be chosen large enough that we can write $f = f_1 + f_2$, where $|f_1| < B$ and the integral of $|f_2|$ is small. Now pick N so big that the set of all points which don't have an ergodic average for f almost as big as b by time N (to be called *bad points*) has a probability that is a small fraction of $\frac{(b-a)}{B}$. Let M $\gg$ N. Show that there is an $S \subset \{T\omega, T^2\omega, \ldots, T^M\omega\}$ such that

(i) the average of f over S is not much smaller than b, and
(ii) only bad points and points among the last N points are outside S. *Hint: express S as a disjoint union of intervals for which the average of f is not much less than b.*

Now f_1 summed over the points in $\{T\omega, T^2\omega, \ldots, T^M\omega\}$ that are outside S is usually a small fraction of $(b - a)M$, because the terms are bounded above by B and the number of bad points is usually a small fraction of $\frac{b-a}{B}$. $|f_2|$ summed over these same points is usually a small fraction of M, since $\int |f_2|$ is a small fraction of M.

This proves that the average of all the points in the interval is usually much closer to b than to a. A similar argument shows that the average is usually much closer to a than to b, leading to a contradiction.

Sketch of proof.

203. Exercise. We may, without loss of generality, assume that T is ergodic. ●

For $\omega \in \Omega$, let $l(\omega) = \liminf_{N\to\infty} \frac{1}{N} \sum_{n=1}^{N} f(T^i\omega)$, $u(\omega) = \limsup_{N\to\infty} \frac{1}{N} \sum_{n=1}^{N} f(T^i\omega)$.

204. Exercise. Show that l and u are measurable and T-invariant. ●

Hence by the ergodicity assumption, l and u take on constant values a and b, respectively. Plainly $a \le b$; we must show that equality holds. Assume for a contradiction that $a < b$.

Choose B so large that one has a decomposition $f = f_1 + f_2$, where $|f_1(\omega)| \le B$ for all ω and $\int |f_2|\, d\mu < \frac{b-a}{64}$. Now choose N so large that the set of all points which don't have an ergodic average for f of at least $b - \frac{b-a}{4}$ by time N (to be called *bad points*[34]) has measure at most $\frac{b-a}{96B}$. Next, pick $M > \frac{24BN}{b-a}$ and let S be the base of a Rohlin tower of height M having an error set of measure smaller than $\frac{1}{2}$ (so that, in particular, $\mu(S) > \frac{1}{2M}$).

---

[34] If it isn't clear what the bad points are, the corresponding set of "good points" is $G = \{\omega : \text{there exists } n \le N \text{ such that } \frac{1}{n} \sum_{i=1}^{n} f(T^i\omega) > b - \delta\}$.

For $\omega \in S$, put $I_\omega = \{\omega, T\omega, T^2\omega, \ldots, T^{M-1}\omega\}$. Let $r(\omega)$ be the number of bad points in I_ω and let $j(w) = \sum_{x \in I_\omega} |f_2(x)|$.

205. Exercise. Show that $\int_S r(\omega)\, d\mu(\omega) \leq \frac{b-a}{96B}$ and $\int_S j(\omega)\, d\mu(\omega) \leq \frac{b-a}{64}$.

                                                    ●

From the previous exercise, we get that the set $D_1 = \{\omega \in S : r(\omega) > \frac{b-a}{12B}M\}$ satisfies $\mu(D_1) < \frac{1}{8M}$ and the set $\{\omega \in S : j(\omega) > \frac{b-a}{b}M\}$ satisfies $\mu(D_2) < \frac{1}{8M}$. Hence letting $S' = S \setminus (D_1 \cup D_2)$, one has $\mu(S') > \frac{1}{2}\mu(S)$.

206. Exercise. Show that, for any $\omega \in S$, there exists a set $E_\omega \subset I_\omega$ such that:

(i) if $E_\omega \neq \emptyset$ then $\frac{1}{|E_\omega|} \sum_{x \in E_\omega} f_1(x) > b - \frac{b-a}{4}$;

(ii) if $x \in I_\omega \setminus E_\omega$ then either $x \in \{T^{M-N}\omega, T^{M-N+1}\omega, \ldots, T^{M-1}\omega\}$ or x is bad.

                                                    ●

Conclude[35] that if $\omega \in S'$ then $\frac{1}{M}\sum_{i=0}^{M-1} f(T^i\omega) > \frac{b+a}{2}$. This shows that, for more than half the points ω in the base of the tower, $\sum_{i=0}^{M-1} f(T^i\omega) > \frac{b+a}{2}$. But it's just as easy to show that for more than half the points ω in the base of the tower, $\sum_{i=0}^{M-1} f(T^i\omega) < \frac{b+a}{2}$. This is a contradiction.   □

207. Comment. The Birkhoff ergodic theorem holds for non-invertible systems with σ-finite measure (see e.g. [Walters 2000, Theorem 1.14]), systems with continuous time (see Krengel 1985, p. 10, for discussion), as well as for some systems without an invariant measure. Indeed, a (possibly non-measure preserving) system $(\Omega, \mathcal{A}, \mu, T)$ is said to be *asymptotically mean stationary*, or *AMS*, if $\lim_n \frac{1}{n}\sum_{i=1}^{n} \mu(T^{-i}A)$ exists for every measurable set A. For a development of the theory of such systems, including a relevant extension of the Birkhoff ergodic theorem, see Gray (1988, Chapters 6–8).

208. Exercise. (*Birkhoff ergodic theorem, version 2.*) In the ergodic case of Theorem 202, $\lim_{N \to \infty} \frac{1}{N}\sum_{n=1}^{N} f(T^i\omega) = \int f\, d\mu$. *Hint: first show that*

---

[35] The argument runs as follows. For $\omega \in S'$, one has

$$\sum_{i=0}^{M-1} f(T^i\omega) = \frac{1}{M}\sum_{x \in I_\omega} f_1(x) + \frac{1}{M}\sum_{x \in I_\omega} f_2(x)$$

$$\geq \frac{1}{M}\left(\sum_{x \in E_\omega} f_1(x) - \sum_{x \in I_\omega \setminus E_\omega} |f_1(x)|\right) - \frac{1}{M}\sum_{x \in I_\omega} |f_2(x)|$$

$$\geq \frac{1}{|E_\omega|}\sum_{x \in E_\omega} f_1(x) - \frac{B}{M}\left(N + \frac{b-a}{12B}M\right) - \frac{j(\omega)}{M}$$

$$\geq b - a - \frac{b-a}{4} - \frac{b-a}{24} - \frac{b-a}{12} - \frac{b-a}{8} = \frac{b+a}{2}.$$

the limit is T-invariant, and hence constant by ergodicity. Now use the dominated convergence theorem on f_1 and control the error due to f_2 with Fatou's lemma.

209. Comment. When T is ergodic and f is the indicator function of a measurable set A, the Birkhoff ergodic theorem says that the frequency of times that the orbit of a typical point under T lands in A is $\mu(A)$.

210. Theorem. *(Strong law of large numbers; see e.g. Dudley 2002.) Let $(X_i)_{i=-\infty}^{\infty}$ be an i.i.d. process composed of real-valued X_i. Then $\lim_{N \to \infty} \frac{1}{N} \sum_{i=0}^{N-1} X_i = E(X_0)$ a.e.*

Sketch of proof. Denote by $(Z, \mathcal{B}, \nu)$ the probability space on which the process is defined. Let $n > 0$ be large and let $\Lambda = \{\frac{i}{n} : i \in \mathbf{Z}\}$, which we view as a countable alphabet. Next, define $f : \mathbf{R} \to \Lambda$ by $f(x) = \frac{\lfloor nx \rfloor}{n}$.[36] Then let $Y_i = f \circ X_i$.

211. Exercise. $(Y_i)_{i=-\infty}^{\infty}$ is an independent stationary process on a countable alphabet.   •

Let $(\Omega, \mathcal{A}, \mu, T)$ be the measure-preserving system associated with (Y_i). By Theorem 149, $(\Omega, \mathcal{A}, \mu, T)$ is ergodic. Define a measurable function $g : \Omega \to \mathbf{R}$ by $g\big((x_i)_{i=-\infty}^{\infty}\big) = f(x_0)$.

212. Exercise. Show that $\int g \, d\mu = \sum_{i=-\infty}^{\infty} \frac{i}{n} P(Y_0 = \frac{i}{n}) = E(Y_0)$.   •

Hence by the Birkhoff ergodic theorem, $\lim_{N \to \infty} \frac{1}{N} \sum_{n=1}^{N} g(T^i \omega) = \int g \, d\mu = E(Y_0)$ for a.e. $\omega \in \Omega$.

213. Exercise. Define $\pi : Z \to \Omega$ by $\pi(z) = \big(Y_i(z)\big)_{i=-\infty}^{\infty}$. Show that π is measure-preserving and that $Y_i(z) = g(T^i(\pi(z)))$ a.e. for every $i \in \mathbf{Z}$. Conclude that $\lim_{N \to \infty} \frac{1}{N} \sum_{n=1}^{N} Y_i(z) = E(Y_0)$ a.e. and use this to complete the proof.   □

214. Comment. The above proof is a bit inefficient, since we are dealing in this book only with processes on countable alphabets. Actually, one can turn real-valued processes into measure-preserving systems just as easily. Briefly, let (X_i) be such a process. We require that the domain space be $(\Omega, \mathcal{A}, \mu)$, where $\Omega = \mathbf{R}^{\mathbf{Z}}$, $\mathcal{A}$ is generated by sets of the form $\{(x_i)_{i=-\infty}^{\infty} : x_j \in (\frac{k}{2^n}, \frac{k+1}{2^n}]\}$, where $k, j \in \mathbf{Z}, n \in \mathbf{N}$, and

$$P(X_{i_1} \in S_1, \ldots, X_{i_k} \in S_k) = \mu\big\{(x_i)_{i=-\infty}^{\infty} \in \Omega : x_{i_1} \in S_1, \ldots, x_{i_k} \in S_k\big\}$$

---

[36] Here $\lfloor y \rfloor$ denotes the greatest integer less than or equal to y.

for all $i_1, \ldots, i_k \in \mathbf{Z}$, where $S_1, \ldots, S_k$ are sets of the form $(\frac{k}{2^n}, \frac{k+1}{2^n}]$. We also require (in order to avoid pathology) that $(\Omega, \mathcal{A}, \mu)$ be a Lebesgue space. One must show that the shift T preserves μ, whereupon one has a measure-preserving system $(\Omega, \mathcal{A}, \mu, T)$.

215. Exercise. Formalize the above and use it to give a more direct proof of the strong law. *Hint: show that an independent real-valued process is ergodic.*

216. Exercise. Show that a stationary process with a countable alphabet is ergodic if and only if the frequency of every word is constant.[37]

2.7. Measure from a monkey sequence

We outline a way to construct a measure-preserving system from a random sequence of letters.

217. Discussion. Put $\Omega = \Lambda^{\mathbf{Z}}$, where Λ is a countable alphabet. Now let an immortal monkey type any infinite sequence $\lambda_1, \lambda_2, \ldots$ of letters from Λ.[38] We can use this sequence to construct a shift-invariant measure on Ω as follows.

Select increasing integers m_i in such a way that for every word w of length 1, the frequency of w in the sequence of finite words $\lambda_1 \lambda_2 \cdots \lambda_{m_i}$ converges.[39]

Now take a subsequence of (m_i), call it (n_i), in such a way that for every word w of length 2, the frequency of w in the sequence of finite words $\lambda_1 \lambda_2 \cdots \lambda_{n_i}$ converges.[40]

Continue taking subsequences, using the standard diagonal argument to converge to a stable sequence (r_i) in the end, having the property that for every $k \in \mathbf{N}$ and word of length k, the frequency of w in the sequence of finite words $\lambda_1 \lambda_2 \cdots \lambda_{r_i}$ converges.[41]

---

[37] Translation: let $(X_i)_{i=-\infty}^{\infty}$ be a stationary process on a countable alphabet Λ. Form the associated measure-preserving system $(\Omega, \mathcal{A}, \mu, T)$, where $\Omega = \Lambda^{\mathbf{Z}}$, etc. $(\Omega, \mathcal{A}, \mu, T)$ is ergodic if and only if for every $k \in \mathbf{N}$ and every k-tuple $w = (\lambda_0, \lambda_1, \ldots, \lambda_{k-1}) \in \Lambda^k$, there exists a number f_w (the frequency of the word w) such that for a.e. $(x_i)_{i=-\infty}^{\infty} \in \Omega$, $\lim_{N \to \infty} \frac{|\{n \in [-N, N] : x_n = \lambda_0, x_{n+1} = \lambda_1, \ldots, x_{n+k-1} = \lambda_{k-1}\}|}{2N+1} = f_w$. Another way of putting this is to say that with probability 1, $\lim_{N \to \infty} \frac{|\{n \in [-N, N] : X_n = \lambda_0, X_{n+1} = \lambda_1, \ldots, X_{n+k-1} = \lambda_{k-1}\}|}{2N+1} = f_w$. The reader should get used to passing seamlessly back and forth between such interpretations.

[38] In other words, let us consider an arbitrary sequence in Ω.

[39] This means that $p_c = \lim_{i \to \infty} \frac{|\{n \in \{1, \ldots, m_i\} : \lambda_n = c\}|}{m_i}$ exists for $c \in \{0, 1\}$.

[40] This means that $p_{cd} = \lim_{i \to \infty} \frac{|\{n \in \{1, \ldots, n_i - 1\} : \lambda_n = c, \lambda_{n+1} = d\}|}{n_i}$ exists for $c, d \in \{0, 1\}$.

[41] That is, $p_w = \lim_{i \to \infty} \frac{|\{n \in \{1, \ldots, n_i - k+1\} : \lambda_n \lambda_{n+1} \cdots \lambda_{n+k-1} = w\}|}{r_i}$ exists.

218. Exercise. Use the frequencies of the words w to construct a premeasure on the algebra of finite unions of cylinder sets. Show that the measure you get upon applying Carathéodory's theorem is stationary.

219. Definition. We call the above method for extracting a stationary measure from a sequence on a finite alphabet the *monkey method*. Of course, the method doesn't have anything to do with monkeys. (See Furstenberg 1981 for a treatment utilizing the Riesz representation theorem.)

220. Convention. When Λ is an alphabet and $\Omega = \Lambda^{\mathbf{Z}}$, if $x \in \Omega$ then we use the notation x_i for the ith coordinate of x; that is to say, we assume without mention that $x = (x_i)_{i=-\infty}^{\infty}$.

221. Definition. Let Λ be a countable alphabet and let $\Omega = \Lambda^{\mathbf{Z}}$. For a word $x \in \Omega$ and $n \in \mathbf{N}$, write $w_n(x)$ for the word $x_1 x_2 \ldots x_n$. Let $W_n = \Lambda^n$ be the set of words of length n. Suppose $k < m$, w is a word of length k and u is a word of length m. We write

$$rf(w; u) = \frac{\#\{i : 1 \leq i \leq m - k + 1, u_i u_{i+1} \cdots u_{i+k-1} = w\}}{m - k + 1}$$

for the relative frequency with which w occurs as a subword of u. If $B \subset W_k$ we write

$$rf(B; u) = \frac{\#\{i : 1 \leq i \leq m - k + 1, u_i u_{i+1} \cdots u_{i+k-1} \in B\}}{m - k + 1}$$

for the relative frequency with which the members of B occur collectively as subwords of u. If $v \in \Omega$ we write $rf(w; v) = \lim_{n \to \infty} rf(w; w_n(v))$, should this limit exist, and similarly for $rf(B; v)$.

222. Definition. Suppose $(\Omega, \mathcal{A}, \mu, T)$ is a measure-preserving system derived from a stationary process (so that $\Omega = \Lambda^{\mathbf{Z}}$, T is the shift, etc.). For $k \in \mathbf{N}$ and $B \subset W_k$, we let $\varphi(B) = \{x \in \Omega : w_n(x) \in B\}$. When B is a singleton $\{w\}$, we may write $\varphi(w)$ rather than $\varphi(\{w\})$. We say that $x \in \Omega$ is *generic* if for every finite word w, $rf(w, x)$ exists.

223. Comment. In order to create a stationary measure via the monkey method from a generic point, it isn't necessary to pass to subsequences. Note that if μ is the resulting measure, $rf(w, x) = \mu(\varphi(\{w\}))$ for every finite word w. We express this by saying that x is *generic for μ*. It is important to note that the notion of a generic point is not more general than that of one generic for some μ; every generic sequence is generic for its own monkey measure.

2.8. Ergodic decomposition

In this subchapter, we show how to decompose an arbitrary measure-preserving system as an integral of ergodic ones. (This allows one in many applications to restrict attention to ergodic systems without any loss of generality.)

224. Exercise. Let $(X, \mathcal{A}, \mu, T)$ be a measure-preserving system and suppose that $\mathcal{C}$ is dense in the measure algebra.[42] $(X, \mathcal{A}, \mu, T)$ is an ergodic system if and only if for every $C \in \mathcal{C}$ one has $\lim_n \frac{1}{n} \sum_{i=1}^n 1_C(T^i x) = \mu(C)$ a.e. *Hint: for an arbitrary T-invariant $A \in \mathcal{A}$ show that $\int |1_A - \mu(A)| \, d\mu$ is arbitrarily small by approximating A by some $C \in \mathcal{C}$ and using the triangle inequality in the integrand.*

225. Exercise. Suppose $(\Omega, \mathcal{A}, \mu, T)$ is a measure-preserving system derived from a stationary process (so that $\Omega = \Lambda^{\mathbf{Z}}$, T is the shift, etc.). $(X, \mathcal{A}, \mu, T)$ is an ergodic system if and only if for every finite word w one has $\lim_n \frac{1}{n} \sum_{i=1}^n 1_{\varphi(w)}(T^i x) = \mu\big(\varphi(w)\big)$ in measure. *Hint: use the Birkhoff theorem and Exercise 224, keeping in mind that if a sequence converges a.e. to f and in measure to g then $f = g$.*

226. Theorem. *Suppose $(\Omega, \mathcal{A}, \mu, T)$ is a measure-preserving system derived from a stationary process (so that $\Omega = \Lambda^{\mathbf{Z}}$, T is the shift, etc.) and $x \in \Omega$ is generic for μ and has the following property $\mathcal{P}(x)$:*

$\mathcal{P}(x)$: For any $k \in \mathbf{N}$ and any word $w \in W_k$ there exists $c(w, x)$ such that for every $\epsilon > 0$ there is an M_0 such that if $m > M_0$ there exists some N_0 such that for all $n > N_0$, if we let $B = B(m, w, x) = \{u \in W_m : |rf(w; u) - c(w, x)| \geq \epsilon\}$ then $rf\big(B, w_n(x)\big) < \epsilon$.

Then $(\Omega, \mathcal{A}, \mu, T)$ is ergodic.

227. Comment. We call $\mathcal{P}(x)$ the "little, middle, big" condition. You can think of it as saying that for every little (word w) there is a middle (size M) such that for every big (initial word u of x), most middle-sized subwords of big have approximately the same frequency of occurrences of little.

Idea of proof. Let x be generic and suppose $\mathcal{P}(x)$ is satisfied. The first thing to realize is that for every finite word w, $c(w, x) = rf(w; x)$ ($rf(w; x)$ exists because x is generic, and $c(w, x)$ is the only viable candidate). Our aim is to show that for every finite word w,

$$\lim_n \frac{1}{n} \sum_{i=1}^n 1_{\varphi(w)}(T^i x) = \mu\big(\varphi(w)\big)$$

---

[42] In other words, for any $\epsilon > 0$ and any $A \in \mathcal{A}$ there is some $C \in \mathcal{C}$ with $\mu(A \triangle C) < \epsilon$.

in measure, which will complete the proof by Exercise 225. Here's an idea of how to do that. First note that for any fixed m, if n is large enough then $w_n(x)$ essentially tells you the distribution of words of length m. Using this fact, translate the above sentence about convergence in measure into a sentence about $w_n(x)$ and you'll see that what you get is just $\mathcal{P}(x)$.

Sketch of proof. Fix a finite word w. It's sufficient to show:

(*) For every $\epsilon > 0$ there exists M_0 such that if $m > M_0$ then

$$\mu\left(\left\{x : \left|\frac{1}{m}\sum_{i=1}^{m} 1_{\varphi(w)}(T^i x) - \mu\big(\varphi(w)\big)\right| \geq 2\epsilon\right\}\right) < \epsilon.$$

Let $\epsilon > 0$. Choose M_0 as in condition $\mathcal{P}(x)$ and bigger than twice the length of w. Now let $m > M_0$. According to condition $\mathcal{P}(x)$, for all sufficiently large n, if we let $B = \{u \in W_m : |rf(w; u) - c(w, x)| \geq \epsilon\}$ then $rf\big(B, w_n(x)\big) < \epsilon$. This implies, in turn, that $rf(B, x) \leq \epsilon$. But x is generic for μ, hence $\mu\big(\varphi(B)\big) = rf(B, x) \leq \epsilon$.

228. Exercise. Note that $\mu\big(\varphi(B)\big) = c(w, x)$, and use this to show that

$$\left\{x : \left|\frac{1}{m}\sum_{i=1}^{m} 1_{\varphi(w)}(T^i x) - \mu\big(\varphi(w)\big)\right| \geq 2\epsilon\right\} \subset \varphi(B). \qquad \square$$

229. Theorem. *Suppose $(\Omega, \mathcal{A}, \mu, T)$ is a measure-preserving system derived from a stationary process (so that $\Omega = \Lambda^{\mathbf{Z}}$, T is the shift, etc.). For a.e. x with respect to μ, $\mathcal{P}(x)$ holds.*

Idea of proof. The Birkhoff ergodic theorem says that for any "little" (word) the frequency of times little occurs converges to a (not necessarily constant) limit c, so by some time "middle" it will probably be very close to c. Again by the Birkhoff ergodic theorem, now for most initial words "big", most middle-sized words in that word exhibit copies of little with frequency near c.

Sketch of proof. Now for the details.

Fix $\epsilon > 0$ and a finite word w. For $m \in \mathbf{N}$ let

$B^m = \big\{y \in \Omega : \text{there exists } m' \geq m \text{ such that } |rf\big(w; w_{m'}(y)\big) - rf(w; y)| > \epsilon\big\}.$

Notice that $B^{m+1} \subset B^m$.

230. Exercise. Use the Birkhoff ergodic theorem applied to the function $1_{\varphi(w)}$ to show that $\lim_{m\to\infty} \mu(B^m) = 0$. $\qquad\bullet$

Let now $l_m(x) = \lim_{n\to\infty} \frac{1}{n}\sum_{i=1}^{n} 1_{B^m}(T^i x)$, which exists a.e. by Birkhoff.

231. Exercise. Use the dominated convergence theorem to show $\int l_m \, d\mu = \mu(B^m)$.

Now put $E^m = \{x : l(x) \geq \epsilon\}$.

232. Exercise. Show that $E^{m+1} \subset E^m$. Show also that if $\delta > 0$ and $\mu(B^m) < \epsilon\delta$ then $\mu(E^m) < \delta$. Conclude that $\lim_{m \to \infty} \mu(E^m) = 0$.

Let now x be any generic point in $\left(\bigcap_{m=1}^{\infty} E^m \right)^c$. This x has the property that there exists some M_0 such that $x \notin E^m$ for every $m \geq M_0$, which means that $l_m(x) = \lim_{n \to \infty} \frac{1}{n} \sum_{i=1}^{n} 1_{B^m}(T^i x) < \epsilon$. This implies that there exists an N_0 such that for every $n > N_0$, $\frac{1}{n} \sum_{i=1}^{n} 1_{B^m}(T^i x) < \epsilon$.

233. Exercise. Show that $T^i x \in B^m$ implies that $T^i x \in \varphi\big(B(m, w, x)\big)$. (Warning: this does not mean that $B^m \subset \varphi\big(B(m, w, x)\big)$.) Conclude that for every $n > N_0$, $\frac{1}{n} \sum_{i=1}^{n} 1_{\varphi(B(m,w,x))}(T^i x) < \epsilon$, hence that $rf\big(B(m, w, x), w_n(x)\big) < \epsilon$.

What we have shown is that for a given finite word w and a given $\epsilon > 0$, for a.e. x, x is generic and there is an M_0 such that if $m > M_0$ there exists some N_0 such that for all $n > N_0$, $rf\big(B(m, w, x), w_n(x)\big) < \epsilon$. Denote this good set of x by $G(w, \epsilon)$.

234. Exercise. Show that any $x \in G = \bigcap_{w,h} G(w, \frac{1}{h})$ is generic and satisfies $\mathcal{P}(x)$. □

235. Theorem. *(Ergodic decomposition; Rohlin.) Suppose $(\Omega, \mathcal{A}, \mu, T)$ is a measure-preserving system derived from a stationary process (so that $\Omega = \Lambda^{\mathbb{Z}}$, T is the shift, etc.). There exists an a.e. defined function $x \to \mu_x$ taking Ω to probability measures on $(\Omega, \mathcal{A})$ such that:*

(1) μ_x is ergodic a.e.;
(2) for any μ-integrable $f : \Omega \to \mathbf{R}$, f is integrable with respect to μ_x a.e. and

$$\int f \, d\mu = \int \left(\int f \, d\mu_x \right) d\mu(x).$$

First proof of Theorem. 235. We'll give two proofs of this theorem. This is the first. For any generic x such that $\mathcal{P}(x)$ holds, let μ_x be its monkey measure.[43] Then by Theorem 229 μ_x is defined a.e. and by Theorem 226 μ_x is ergodic whenever it is defined.

---

[43] That is, the measure v constructed in Theorem 226. Notice that it is unique, by Comment 223.

236. Exercise. Recall the discussion in 217. Show that if C is a cylinder set then $\frac{1}{n}\sum_{i=1}^{n}1_C(T^i x) \to \mu_x(C)$ a.e. Conclude that $\mu(C) = \int \mu_x(C)\,d\mu(x)$. Use this fact to show that the conclusion to the theorem holds whenever f is a finite linear combination of cylinder set indicator functions.   ●

In order to get the result in full from the foregoing exercise, we have to be able to deal with the error function that arises when you approximate a general f by a finite linear combination of cylinder set indicator functions.

237. Claim. Let $\delta > 0$. If $\mu(A) < \delta^2$ then $\mu(\{x : \mu_x(A) > \delta\}) \leq \delta$.

Assume for contradiction that the claim fails, that is, $\mu_x(A) > \delta$ for all $x \in B$, where $\mu(B) > \delta$. Choose cylinder sets $(C_i)_{i=1}^{\infty}$ with $A \subset \bigcup_{i=1}^{\infty} C_i$ and $\sum_{i=1}^{\infty}\mu(C_i) < \delta^2$.

238. Exercise. Show that there exist $n \in \mathbf{N}$ and $B' \subset B$ such that $\mu(B') > \delta$ and $\sum_{i=1}^{n}\mu_x(C_i) > \delta$ for $x \in B'$. Use Exercise 236 to conclude that $\sum_{i=1}^{n}\mu(C_i) > \delta$, a contradiction establishing the claim.

239. Exercise. If $\delta > 0$ and if $f \geq 0$ with $\int f\,d\mu < \delta^2$ then $\int \left(\int f\,d\mu_x\right) d\mu(x) \leq \delta^2$. *Hint: virtually the same proof; pick (C_i) this time with $f \leq \sum_{i=1}^{\infty}a_i 1_{C_i}$ and $\sum_{i=1}^{\infty}a_i\mu(C_i) < \delta^2$; use the monotone convergence theorem.*   ●

The foregoing exercise makes it easy to deal with the error you get when approximating by a linear combination of cylinder set indicator functions. Anyway this does it for non-negative functions and any integrable function is the difference of two non-negative functions – we'll leave verification of the final details to the reader.   □

Our second proof of Theorem 235 requires a bit of preparation, so will be deferred for now.

240. Exercise. (*Birkhoff ergodic theorem, version 3.*) In the general case of Theorem 202, $\lim_{N\to\infty}\frac{1}{N}\sum_{n=1}^{N}f(T^i\omega) = E(f|\mathcal{B})$, where $\mathcal{B}$ is the σ-algebra generated by the T-invariant sets. *Hint: first show that the limit h is T-invariant, hence $\mathcal{B}$-measurable. Then show (cf. Exercise 208) that $\int_B h\,d\mu = \int_B f\,d\mu$ for every $B \in \mathcal{B}$.*

2.9. Ergodic theory on L^2

One of the themes of this book and of the methodology it champions is that you can do a lot of involved ergodic theory without assuming a lot of structure. In the next few sections we'll talk a little bit, for the sake of completeness, about perhaps the most basic unit of additional structure one might add under a different philosophy: $L^2(\Omega, \mathcal{A}, \mu)$. Nothing we say here will be needed until the optional last chapter, so it may be skipped without dire consequence. We do assume a basic familiarity with Hilbert spaces.

241. Discussion. If $(\Omega, \mathcal{A}, \mu)$ is a probability space then the set $L^2(\Omega, \mathcal{A}, \mu)$ of complex-valued square integrable functions[44] f on Ω is a Hilbert space. The inner product on this space is given by $\langle f, g \rangle = \int f \overline{g} \, d\mu$. The important thing to note, from the standpoint of ergodic theory, is that if $T : \Omega \to \Omega$ is an invertible measure-preserving transformation then the map $f \to Tf$, where $Tf(x) = f(Tx)$, is unitary on $L^2(\Omega, \mathcal{A}, \mu)$.[45]

Let's have a brief look at how the objects we have been studying relate to L^2:

242. Theorem. *The map* $f \to E(f|\mathcal{B})$ *on* $L^2(\Omega, \mathcal{A}, \mu)$ *is the orthogonal projection onto the subspace of* $\mathcal{B}$ *measurable, square integrable functions.*

Sketch of proof.

243. Exercise. Fix $f \in L^2(\Omega, \mathcal{A}, \mu)$. It suffices to show that for an arbitrary $\mathcal{B}$-measurable, square integrable function g, $||f - g|| \geq ||f - E(f|\mathcal{B})||$.   ●

Fix such a g. We need one more thing:

244. Exercise. Let $(X, \mathcal{C}, \nu)$ be an arbitrary probability space and $f \in L^2(X)$. Show that $\inf_{c \in \mathbf{R}} ||f - c|| = ||f - \int f \, d\nu||$.   ●

Now we have

$$||f - g|| = \int |f - g|^2 \, d\mu$$
$$= \int \left(\int |f - g|^2 \, d\mu_x \right) d\mu(x)$$
$$\leq \int \left(\int \left| f - \int f \, d\mu_x \right|^2 d\mu_x \right) d\mu(x)$$
$$= \int \left| f - \int f \, d\mu_x \right|^2 d\mu(x)$$
$$= \int |f - E(f|\mathcal{B})|^2 \, d\mu(x) = ||f - E(f|\mathcal{B})||. \qquad (1)$$

$\qquad\qquad\qquad\qquad\qquad\qquad\qquad\qquad\qquad\qquad\qquad\qquad\qquad\qquad\qquad$ □

We can now give an alternate proof of the result in Exercise 240.

---

[44] That is, the functions f such that $\int |f|^2 \, d\mu < \infty$.

[45] That is, T is a linear operator satisfying $\langle Tf, Tg \rangle = \langle f, g \rangle$ for all $f, g \in L^2(\Omega, \mathcal{A}, \mu)$.

245. Corollary. *In the Birkhoff ergodic theorem, one has* $\lim_{n\to\infty} \frac{1}{n} \sum_{i=1}^{n}$
$f(T^i x) = E(f|\mathcal{B})(x)$ *a.e., where* $\mathcal{B}$ *is the* σ*-algebra of sets that are*
T*-invariant mod 0.*

Sketch of proof.

246. Exercise. In general, for $f \in L^2((\Omega, \mathcal{A}, \mu))$ let $Pf = \lim_{N\to\infty}$
$\frac{1}{N} \sum_{n=1}^{N} f(T^i \omega)$. Show that P is the orthogonal projection onto the closure
I of the space of T-invariant functions. *Hint: show first that* Pf *is always*
T*-invariant. Then show that* $g \in I^{\perp}$ *implies* $Pg \in I^{\perp}$.   •

Invoking Theorem 242 now finishes the proof.   □

2.10. Conditional expectation of a measure

Earlier we showed how to decompose a system into ergodic components. Here
we give a more general construction, decomposing a measure over an arbitrary
σ-algebra.

247. Theorem. *(Rohlin 1952.) Suppose* $(\Omega, \mathcal{A}, \mu, T)$ *is a measure-preserving*
system derived from a stationary process (so that $\Omega = \Lambda^{\mathbf{Z}}$, T *is the shift, etc.)*
and let $\mathcal{B} \subset \mathcal{A}$ *be a sub-σ-algebra that is complete with respect to* μ. *There*
exists an a.e. defined function $x \to \mu_x$ *taking* Ω *to probability measures on*
$(\Omega, \mathcal{A})$ *such that:*

(1) *for any* μ*-integrable* f, *the function* $g(x) = \int f \, d\mu_x$ *is* $\mathcal{B}$*-measurable;*
(2) *for any* μ*-integrable* $f : \Omega \to \mathbf{R}$, f *is integrable with respect to* μ_x *a.e.*
    and for any $B \in \mathcal{B}$,

$$\int_B f \, d\mu = \int_B \left(\int f \, d\mu_x \right) d\mu(x);$$

(3) *if* $\mathcal{B}$ *is* T*-invariant mod 0 then for a.e.* x, T *maps* $(\Omega, \mathcal{A}, \mu_x)$ *to*
    $(\Omega, \mathcal{A}, \mu_{Tx})$ *in a measure-preserving fashion.*

248. Comment. It is intuitively useful to employ a doubly indexed nota-
tion. For example, let $E_{\mu|\mathcal{B}}(A, x) = \mu_x(A)$. Now for fixed x, the map
$A \to E_{\mu|\mathcal{B}}(A, x)$ gives back the measure μ_x and for fixed A the map
$x \to E_{\mu|\mathcal{B}}(A, x)$ gives back a version of the conditional expectation of 1_A.
One of the authors prefers this notation and feels that it appears to indicate a
(false) proof of this theorem. Investigation of what this idea is, and why it fails,
can help one to appreciate the subtlety of the result.

The idea is that you just pick a version of $P(A|\mathcal{B})$ for every measurable
set A, write the value of this at x as $E_{\mu|\mathcal{B}}(A, x)$, and hope now that for

a.e. x the map $A \to E_{\mu|\mathcal{B}}(A, x)$ gives you a measure. Things look promising enough from the standpoint that $P(\bigcup_{i=1}^{\infty} A_i | \mathcal{B})(x) = \sum_{i=1}^{\infty} P(A_i | \mathcal{B})(x)$ certainly holds (for a.e. x) for pairwise disjoint sequences $(A_i)_{i=1}^{\infty}$. But, unfortunately, there are uncountably many choices for the sequence $(A_i)_{i=1}^{\infty}$. That is why we must restrict attention to a countable dense set of sets (namely the set of finite unions of cylinder sets).

Sketch of proof. We construct the family of measure μ_x as follows. For every cylinder set C, pick a version of $E(1_C | \mathcal{B})$ and call it, say, f_C. Now for a given $x \in \Omega$, let $p_x(C) = f_C(x)$.

249. Exercise. Show that for a.e. x with respect to μ, p_x extends uniquely to a premeasure on the algebra of finite unions of cylinder sets. *Hint: use Exercise 124.* •

Now by Caratheódory's extension theorem, p_x extends to a measure μ_x a.e.

250. Exercise. Show that for $A \in \mathcal{A}$, $x \to \mu_x(A)$ is $\mathcal{B}$-measurable. *Hint: clear for cylinder sets; show that the set of A satisfying the conclusion forms a complete σ-algebra.* •

By the foregoing exercise, $x \to \int f \, d\mu_x$ will be $\mathcal{B}$-measurable for f a finite linear combination of indicator functions. To get to non-negative integrable functions, we just need monotone limits. Accordingly:

251. Exercise. Let $(g_i)_{i=1}^{\infty}$ be a non-decreasing sequence of $\mathcal{B}$-measurable functions and suppose $g = \lim_i g_i$ is integrable. Show that g is $\mathcal{B}$-measurable. Use this fact to prove (1). •

To prove (2), the reader who has been paying attention should now have no trouble.[46] As for (3), we won't be using it so we leave it as an exercise. □

252. Exercise. Let $(\Omega, \mathcal{A}, \mu)$ be a Lebesgue space, $\mathcal{B} \subset \mathcal{A}$ a sub-σ-algebra and let $\{\mu_x : x \in X\}$ be the decomposition of μ over $\mathcal{B}$. Show that if g is a $\mathcal{B}$-measurable function on Ω then for a.e. x (with respect to μ), g is constant a.e. (with respect to μ_x).

Second proof of Theorem. 235. It suffices to show that if $\mathcal{B}$ is the σ-algebra of T-invariant sets then μ_x is ergodic a.e.

---

[46] The formula clearly works when f is an indicator function of a cylinder set. Now either approximate a general f by a finite linear combination of cylinder set indicator functions and control the error effect, as in Exercise 239, or first establish that the formula holds for general indicator functions and use monotone convergence (notice that a use of monotone convergence was buried in the error analysis of the first method anyway). The choice of method and details are left to the reader.

253. Comment. At first one would think that the ergodicity of μ_x is trivial because for any A in the invariant algebra, $\mu_x(A) = P(A|\mathcal{B})(x) = 1_A$ a.e. and hence only takes on the values 0 and 1 a.e. One now wants to apply Exercise 242, but the problem is that for every A one can get an exceptional null set of xs for which it doesn't work, and there are uncountably many choices for A. Recall we drew attention to a very similar difficulty in Comment 248. The solution here is similar to the solution there: we restrict attention to a dense set via employment of Exercise 224.

We apply Exercise 224 as follows. Let C be the countable algebra generated by cylinder sets. By Corollary 245 and the proof of Theorem 247, for $C \in \mathcal{C}$,

$$\lim_n \frac{1}{n} \sum_{i=1}^n 1_C(T^i x) = E(1_C|\mathcal{B})(x) = \mu_x(C)$$

a.e. with respect to μ, and hence, for a.e. x, a.e. with respect to μ_x. For such x, μ_x is an ergodic measure by Exercise 224. $\qquad\square$

2.11. Subsequential limits, extended monkey method

We give some refinements of the monkey method introduced earlier.

254. Definition. Let Λ be a countable alphabet and let $W = \Lambda^i$ be the set of words of length i. A *cylinder set* is a subset of W obtained by specifying values for finitely many coordinates.[47] If C is a cylinder set and the ith coordinate is unspecified, the *right shift* of C is the cylinder set C' you get by making all the same specifications one coordinate to the right.[48] *Left shift* is defined similarly.

255. Definition. A measure p on W is said to be *stationary* if for every cylinder set C and one of its (left or right) shifts C', $p(C) = p(C')$.

256. Comment. So p is stationary if $p(cat**) = p(*cat*) = p(**cat)$ and similarly for all cats. Here $*$ is a standard wildcard, so that for example $cat** = \{\lambda_1\lambda_2\lambda_3\lambda_4\lambda_5 : \lambda_1 = c, \lambda_2 = a, \lambda_3 = t\}$.

257. Exercise. Let p be a measure on W and suppose that $p(*\lambda_1\lambda_2\cdots\lambda_{i-1}) = p(\lambda_1\lambda_2\cdots\lambda_{i-1}*)$ for all $\lambda_1, \lambda_2, \cdots, \lambda_{i-1}$. Show that p is stationary.

258. Comment. If p is a measure on words of length i and $j < i$ then p induces a measure $p^{(j)}$ on words of length j by the rule $p^{(j)}(cat) = p(cat**)$.

---

[47] Hence the general cylinder set has the form $C = \{\lambda_1\lambda_2\cdots\lambda_i : \lambda_{j_k} = l_k, 1 \le k \le t\}$, where $1 \le j_1 < j_2 < \cdots < j_t \le i$.

[48] That is, $C' = \{\lambda_1\lambda_2\cdots\lambda_i : \lambda_{j_k+1} = l_k, 1 \le k \le t\}$, where $1 \le j_1 < j_2 < \cdots < j_t \le i - 1$.

(Here of course we are taking $i = 5$ and $j = 3$.)[49] The reader should check that this also works when p is a measure on $\Lambda^{\mathbf{N}}$.

259. Definition. Let p_i be a sequence of measures on words of length n, and let p be such as well. We say that p_i *converges to* p if for every word w of length n, $\lim_i p_i(w) = p(w)$.

260. Exercise. For every $i \in \mathbf{N}$, let p_i be a measure on words of length i. We are not assuming that the p_i are stationary. Show that there exists a subsequence such that, along this subsequence, $\lim_i p_i^{(j)} = q_j$ exists for all $j \in \mathbf{N}$. *Hint: use a standard diagonal argument.* Finally show that there is a unique measure q on $\Lambda^{\mathbf{N}}$ such that $q^{(j)} = q_j$ for all $j \in \mathbf{N}$.

261. Definition. The limiting measure q obtained in the foregoing exercise is called a *subsequential limit* of the sequence $(p_i)_{i=1}^{\infty}$.

262. Comment. A subsequential limit needn't be stationary if the approximating measures aren't.

263. Definition. A *stochastic process* is an infinite sequence of random variables $(X_i)_{i=1}^{\infty}$ or $(X_i)_{i=-\infty}^{\infty}$ into an alphabet Λ.

264. Comment. The modifier "stochastic" here mostly just serves to remind the reader that the process needn't be stationary. A measure on (doubly or singly) infinite words determines a stochastic process in the obvious way: for an infinite word $y = (y_i)$, just let $X_i(y) = y_i$.

265. Discussion. We've already exhibited a way to extract a stationary process from a sequence of letters. You can also extract a stationary process from a sequence of measures on finite words of increasing length, or from a sequence of measures on infinite sequences. Here's how it's done. Let $(p_i)_{i=1}^{\infty}$ be measures on words of lengths $(L_i)_{i=1}^{\infty}$, respectively, where $L_i \to \infty$. By passing to a subsequence if necessary, you can assume $L_i > 100^i$ for all i. For each i, derive from p_i a "nearly stationary measure" v_i on words of length i by:

$$v_i(a_1 a_2 \dots a_i) = \frac{1}{L_i - i + 1} \sum_{j=1}^{L_i - i + 1} p_i\Big(\big\{ x_1 x_2 \dots x_{L_i} : x_j x_{j+1} \dots x_{j+i-1}$$

$$= a_1 a_2 \dots a_i \big\}\Big).$$

Finally let v be any subsequential limit of $(v_i)_{i=1}^{\infty}$.

---

[49] More generally, $p^{(j)}(\lambda_1 \lambda_2 \cdots \lambda_j) = p(\lambda_1 \lambda_2 \cdots \lambda_j * * \cdots *)$.

266. Definition. Any measure ν so constructed is called a *measure obtained by the extended monkey method.*

267. Exercise. Prove that a measure obtained by the extended monkey method is stationary.

268. Discussion. It's useful to compare subsequential limits with measures obtained by the extended monkey method. Subsequential limits have the advantage of maintaining local behavior. For example, a measure ν obtained from the extended monkey method may, restricted to the first 10 coordinates, look nothing at all like any of the original measures p_i so restricted. On the other hand any ν obtained via the extended monkey method is stationary; straight subsequential limits of non-stationary p_i needn't be.

Of course, if the p_i are stationary, any subsequential limit of them is necessarily stationary, and there is no advantage to using the extended monkey method.

3

Martingales and coupling

3.1. Martingales

We give a brief introduction to martingales. Those not interested may skip.

269. Definition. A sequence $(X_n)_{n=0}^{\infty}$ of real-valued random variables is called a *martingale* if for every $n \geq 1$, $E(X_n|X_0, X_1, \ldots, X_{n-1}) = X_{n-1}$. A *backward martingale* satisfies $E(X_n|X_{n+1}, X_{n+2}, \ldots) = X_{n+1}$. A martingale $(X_n)_{n=0}^{\infty}$ is *bounded* if there is a constant M such that $|X_n| \leq M$ a.e. for all n.

270. Example. Let X_n be the net winnings (negative values interpreted as losses) that a gambler has obtained after n rounds of an even game of chance. For example if a round consists in flipping a fair coin, and the gambler wins a dollar for heads and loses a dollar for tails, and X_n is the amount he has won after n rounds, then $(X_i)_{i=1}^{\infty}$ is a martingale.

271. Definition. The martingale $(X_i)_{i=1}^{\infty}$ of the previous example is called an *unbiased random walk*. A *biased random walk* is the same thing except one uses a biased coin.

272. Definition. A *stopping time* for a martingale $(X_i)_{i=1}^{\infty}$ is a random variable T taking values in $\mathbf{N}$ such that for all $k \in \mathbf{N}$, $T^{-1}(k)$ is in the σ-algebra generated by $X_1, \ldots, X_k$. An *STRV* (for *stopping time random variable*) is a random variable of the form $Y = X_T$, where $(X_i)_{i=1}^{\infty}$ is a martingale and T is a stopping time for $(X_i)_{i=1}^{\infty}$.

273. Comment. A stopping time can be thought of as a rule for stopping that depends only on information you know from looking at the past relative to the time you stop (and which does stop, with probability 1). So, for example, if you have a prophet who can tell you to stop after n rounds in all those instances where you are to lose round $n + 1$, you might make a lot of money under his guidance, but you aren't utilizing a stopping time.

If a stopping time is bounded, we interpret this as saying that you always stop by time n for some n. The STRV associated with a stopping time is the random variable that takes on the value of the martingale when it stops.

274. Definition. We say that a stopping time *makes money* if $E(Y) > E(X_0)$, where Y is the associated STRV.

275. Example. In the coin tossing example, the stopping time specified by "stop as soon as you are ahead by 10^6 dollars" makes money. Indeed, it makes exactly a million dollars, with probability 1.

The reader is advised not to get too excited about this opportunity. The reason for this is that to employ it, you need not only infinite time, but an infinite bankroll as well. To wit:

276. Exercise. Show that in the coin tossing example, the stopping time specified by "stop as soon as you are ahead by 10^6 dollars or at time 10^{10^6}, whichever comes first" does not make money. *Hint: use induction.*

277. Exercise. Show that in the coin tossing example, the stopping time specified by "stop as soon as you are ahead by 10^6 dollars or behind by 10^{10^6} dollars, whichever comes first" does not make money. *Hint: for $n \in \mathbf{N}$, let Y_n be the STRV coming from the stopping time specified by "stop as soon as you are ahead by 10^6 dollars or behind by 10^{10^6} dollars or by time n, whichever comes first". Show that $Y_n \to Y$ a.e.; use dominated convergence.*

More generally we have the following two theorems. To prove these theorems, just make the appropriate modifications to the foregoing exercises.

278. Theorem. *(Bounded time theorem.) An STRV Y coming from a bounded stopping time satisfies $E(Y|X_0) = X_0$. In particular, bounded stopping times don't make money.*

(Bounded money theorem.) In a bounded martingale, stopping times don't make money. Indeed, for an STRV Y coming from a bounded martingale, $E(Y|X_0) = X_0$.

279. Exercise. Prove the bounded time theorem and the bounded money theorem.

280. Example. Consider the coin tossing martingale, but start with L dollars and use the stopping time specified by "stop when you run out of money or when you reach a million dollars". If p if the probability that you hit a million dollars then the bounded money theorem gives $L = E(X_0) = E(Y) = 10^6 p$. That is, $L = \frac{p}{10^6}$.

281. Definition. Let $(X_i)_{i=1}^\infty$ be a one-sided process taking values in some set S, called the set of *states* of the process. A state $s \in S$ is *recurrent* for the process if it occurs with positive probability and it is the case that if it occurs at any time then it will occur again with probability 1.[50] A non-recurrent state that occurs with positive probability is said to be *transient*.

282. Exercise. Show that if s is recurrent and $P(X_i = s) > 0$ then:

(a) $P(X_j = s$ for infinitely many $j > i | X_i = s) = 1$, and
(b) $\sum_{j=i+1}^{\infty} P(X_j = s | X_i = s) = \infty$.

283. Exercise. Use the foregoing example to show that every state is recurrent for an unbiased random walk.

284. Exercise. Show that if s is transient, $P(X_i = s) > 0$ *and* $(X_i)_{i=1}^{\infty}$ is *stationary* then:

(a) $P(X_j = s$ for infinitely many $j > i | X_i = s) = 0$, and
(b) $\sum_{j=i+1}^{\infty} P(X_j = s | X_i = s) < \infty$.

285. Example. Modify the coin tossing example as follows. Use a coin that comes up heads with probability $\frac{1}{3}$. Start with 1 dollar. In each betting round, you bet half your money on heads at fair odds; that is, 2-1. This means you lose your bet if it comes up tails (this has the effect of halving your fortune) and you win double what you bet if it comes up heads (which has the effect of doubling your fortune). If after some number of rounds, m is the number of heads minus the number of tails, your fortune is 2^m. Stop when $m = 1$ or $m = -10^6$. Let p be the probability that m hits 1. Then by the bounded money theorem, $1 = E(X_0) = E(Y) = 2p + \frac{1-p}{2^{10^6}}$.

286. Exercise. Use the foregoing example to show that every state is transient for a biased random walk.

287. Theorem. (*Martingale convergence theorem; see e.g. Dudley 2002, Section 10.5.*) Let $(X_i)_{i=-\infty}^{\infty}$ be a bounded martingale. Then $\lim_{i \to \infty} X_i(x)$ exists a.e.

Idea of proof. Suspend your disbelief and pretend for a moment that the price of some imaginary stock market index IMAG is a bounded martingale (say bounded by 0 and 1). Let $0 < a < b < 1$ and consider the following investment strategy. You sell a million shares of IMAG whenever it has risen to a price $\geq b$ from a price $\leq a$, and buy a million shares whenever it drops to a price $\leq a$ from a price $\geq b$. Now, brokerage fees aside, you will probably get very rich using this strategy provided the price of IMAG crosses the interval

---

[50] To be more precise: $P(X_i = s) > 0$ for some i, and if $P(X_i = s) > 0$ then $P(X_j = s$ for some $j > i | X_i = s) = 1$.

$[a, b]$ infinitely many times. But the probability of this has to be zero, because you can't lose more than $10^6 a$ dollars, and by the bounded money theorem, your expected net gain is 0. Therefore with probability 1 the price of IMAG crosses $[a, b]$ only finitely many times and since this holds for all rational a, b, the price of IMAG must converge to a limit with probability 1.

Sketch of proof. Let $0 < a < b < 1$. In this proof (Z, ν) denotes the domain space of the random variables under consideration. Let $M \in \mathbf{N}$, and let

$$B = \{z : \left(X_i(z)\right)_{i=-\infty}^{\infty} \text{ crosses } (a, b) \text{ at least } 2M \text{ times}\}.$$

We claim that $\nu(B) \leq \frac{2}{M(b-a)}$. Suppose for a contradiction that the claim fails.

288. Exercise. Show that for some $n \in \mathbf{N}$,

$$B_n = \{z : \left(X_i(z)\right)_{i=1}^{n} \text{ crosses } (a, b) \text{ at least } 2M \text{ times}\}$$

satisfies $\mu(B_n) \geq \frac{1}{M(b-a)}$. •

Let $T_1(z)$ be the lesser of n and $\min\{i : X_i(z) \geq b\}$. Now let $T_2(z)$ be the lesser of n and $\min\{i : i > T_1(z) \text{ and } X_i(z) \leq a\}$. Now for $2 < k \leq M$ and k odd, let $T_k(z)$ be the lesser of n and $\min\{i : i > T_{k-1}(z) \text{ and } X_i(z) \geq b\}$. Finally for $2 < k \leq M$ and k even, let $T_k(z)$ be the lesser of n and $\min\{i : i > T_{k-1}(z) \text{ and } X_i(z) \leq a\}$.

Let $Y = X_{T_1} - X_{T_2} + X_{T_3} - X_{T_4} + \cdots + X_{T_{2M-1}} - X_{T_{2M}}$. Now, by the bounded money theorem, $E(Y) = 0$. But for every $z \in B_n$,

$$Y(z) = X_{T_1(z)} - X_{T_2(z)} + X_{T_3(z)} - X_{T_4(z)} + \cdots + X_{T_{2M-1}(z)} - X_{T_{2M}(z)} \geq M(b-a).$$

289. Exercise. Show that $Y(z) \geq b - 1$ except possibly on a set of measure zero. Conclude that $E(Y) \geq b > 0$, a contradiction proving the claim. Now use the truth of the claim to show that $\{z : \left(X_i(z)\right)_{i=-\infty}^{\infty} \text{ crosses } (a, b) \text{ infinitely many times}\}$ is a null set. Finish the proof by considering all pairs (a, b) of rationals. *Question: why are we limiting ourselves to rationals?* □

290. Comment. Bounded backward martingales also converge (by the same argument).

291. Definition. Let $(X_i)_{i=-\infty}^{\infty}$ be a stationary process. The *past* is the σ-algebra generated by $(X_i)_{i<0}$. We will often denote this σ-algebra by *past*.

292. Discussion. So, when the X_i are real valued and we speak of the conditional expectation of the present given the past, we mean $E(X_0|past)$. In the general case, where the X_i may take values in some alphabet Λ, if we speak of the probability of seeing some letter a given the past, we mean $P(X_0 = a|past) = E(1_{X_0^{-1}(a)}|past)$.

This relates to martingales as follows. If you let $Y_i = P(X_0 = a | X_{-1}, X_{-2}, \ldots, X_{-i})$, $(Y_i)_{i=-\infty}^{\infty}$ forms a bounded martingale and must therefore converge a.e. It doesn't take a lot of checking to see that the limit must be $P(X_0 = a | past)$.

3.2. Coupling; the basics

In this subchapter we give basic definitions relating to coupling and introduce simple techniques. (Coupling also goes by the name *joining*: the classical ergodic theory source is Furstenberg (1967), though at a probabilistic level, the idea has been around for a long time.)

293. Definition. Let $(\Omega_i, \mathcal{A}_i)$ be measurable spaces, $i = 1, 2$, and suppose μ is a measure on the product space $(\Omega_1 \times \Omega_2, \mathcal{A}_i \otimes \mathcal{A}_2)$. For $A \in \mathcal{A}_1$, let $\mu_1(A) = \mu(A \times \Omega_2)$ and for $B \in \mathcal{A}_2$, let $\mu_2(B) = \mu(\Omega_1 \times B)$. Then μ_1, μ_2 are called the *marginal probabilities* of μ on the first and second coordinates, respectively, and μ is called a *coupling* of μ_1 and μ_2.

Couplings of more than two measures are defined analogously. For example, a coupling of three measures is a measure on the triple product space whose marginals give back the original measures.

294. Comment. In Section 152 it was pointed out that when an alphabet Λ is fixed, a stationary process $(X_i)_{i=-\infty}^{\infty}$ is characterized by the measure μ it puts on $\Omega = \Lambda^{\mathbf{Z}}$. In a similar fashion, one may identify a one-sided process $(X_i)_{i=1}^{\infty}$ with the measure it induces on $\Lambda^{\mathbf{N}}$. Accordingly, we shall often speak of a coupling between two processes, rather than two measures. Of course, we simply mean the corresponding measures. Similarly, we may speak of coupling two finite sets of random variables $(X_i^{(j)})_{i=1}^{n}$, $j = 1, 2$. What we mean in this instance is the coupling of the corresponding measures that are induced on Λ^n.[51]

295. Definition. Given probability spaces $(\Omega_i, \mathcal{A}_i, \mu_i)$, $i = 1, 2$, the *independent coupling* of μ_1 and μ_2 is the product measure $\mu = \mu_1 \times \mu_2$.

296. Comment. The independent coupling always exists, though there may be others. Often the coupling that maximizes the measure of the diagonal is of interest.

297. Example. Let μ and ν be measures on $\{0, 1\}$ defined by

$$\mu(0) = \frac{1}{3}, \quad \mu(1) = \frac{2}{3}, \quad \nu(0) = \frac{1}{4}, \quad \mu(1) = \frac{3}{4}.$$

---

[51] We mean the measures $\mu(\{\lambda_1, \ldots, \lambda_n\}) = P(X_1 = \lambda_1, \ldots, X_n = \lambda_n)$.

The independent coupling is given by

$$P(0,0) = \frac{1}{12}, \quad P(0,1) = \frac{1}{4}, \quad P(1,0) = \frac{1}{6}, \quad P(1,1) = \frac{1}{2}.$$

The coupling that assigns the greatest measure to the diagonal is given by

$$P(0,0) = \frac{1}{4}, \quad P(0,1) = \frac{1}{12}, \quad P(1,0) = 0, \quad P(1,1) = \frac{2}{3}.$$

298. Example. Let $(\Omega, \mathcal{A}, \mu)$ be a probability space. Create a measure ν on the product space $(\Omega^2, \mathcal{A} \otimes \mathcal{A})$ as follows. For $A, B \in \mathcal{A}$ put $\nu(A \times B) = \mu(A \cap B)$. Extend to a measure on $\mathcal{A} \otimes \mathcal{A}$ by using Carathéodory's theorem. ν is a coupling of μ with itself.

299. Definition. ν as defined above is called the *diagonal coupling* of μ with itself.

300. Exercise. Show that the map $\pi : \Omega \to \Omega^2$ defined by $\pi(x) = (x, x)$ is a measurable isomorphism from $(\Omega, \mathcal{A}, \mu)$ to $(\Omega^2, \mathcal{A} \otimes \mathcal{A}, \nu)$, where ν is the diagonal coupling of μ with itself.

301. Example. *Coupling by induction.* Let $(X_i)_{i=1}^{\infty}$ and $(Y_i)_{i=1}^{\infty}$ be two processes; say the random variables take values in a countable alphabet Λ. We outline an inductive technique for coupling these processes together. First, couple X_1 and Y_1 together any way you like. Assume next that $(X_i)_{i=1}^{n}$ and $(Y_i)_{i=1}^{n}$ have been coupled.[52] Pick a conditional probability law for X_{n+1} given $(X_i)_{i=1}^{n}$ and $(Y_i)_{i=1}^{n}$ in such a way that when you integrate that probability law over all $(Y_i)_{i=1}^{n}$ you get the conditional probability of X_{n+1} given $(X_i)_{i=1}^{n}$.[53] Then pick a conditional probability law for Y_{n+1} given $(X_i)_{i=1}^{n}$ and $(Y_i)_{i=1}^{n}$ in such a way that when you integrate that probability law over all $(X_i)_{i=1}^{n}$ you get the conditional probability of Y_{n+1} given $(Y_i)_{i=1}^{n}$.[54] Now simply couple the conditional on X_{n+1} with the conditional on Y_{n+1}.[55]

---

[52] The assumption is that some measure P_n on Λ^{2n} has been chosen such that

$P_n(\{(x_1, \dots, x_n, y_1, \dots, y_n) : x_i = \lambda_i, 1 \le i \le n\}) = P(X_i = \lambda_i, 1 \le i \le n)$ and
$P_n(\{(x_1, \dots, x_n, y_1, \dots, y_n) : y_i = \gamma_i, 1 \le i \le n\}) = P(Y_i = \gamma_i, 1 \le i \le n)$.

[53] That is to say: choose a family of measures $\{\mu_{\lambda_1,\dots,\lambda_n}^{\gamma_1,\dots,\gamma_n} : \lambda_1, \dots, \lambda_n, \gamma_1, \dots, \gamma_n \in \Lambda\}$ on Λ having the property that for every $\lambda_1, \dots, \lambda_n$ and $\lambda \in \Lambda$, one has $\sum_{\gamma_1,\dots,\gamma_n \in \Lambda} \mu_{\lambda_1,\dots,\lambda_n}^{\gamma_1,\dots,\gamma_n}(\lambda) = P(X_{n+1} = \lambda | X_i = \lambda_i, 1 \le i \le n)$.

[54] This time choose a family of measures $\{\nu_{\lambda_1,\dots,\lambda_n}^{\gamma_1,\dots,\gamma_n} : \lambda_1, \dots, \lambda_n, \gamma_1, \dots, \gamma_n \in \Lambda\}$ on Λ having the property that for every $\gamma_1, \dots, \gamma_n$ and $\gamma \in \Lambda$, one has $\sum_{\lambda_1,\dots,\lambda_n \in \Lambda} \mu_{\lambda_1,\dots,\lambda_n}^{\gamma_1,\dots,\gamma_n}(\gamma) = P(Y_{n+1} = \lambda | Y_i = \lambda_i, 1 \le i \le n)$.

[55] That is, for every $\lambda_1, \dots, \lambda_n, \gamma_1, \dots, \gamma_n \in \Lambda$, couple $\mu_{\lambda_1,\dots,\lambda_n}^{\gamma_1,\dots,\gamma_n}$ with $\nu_{\lambda_1,\dots,\lambda_n}^{\gamma_1,\dots,\gamma_n}$ any way you like and call this coupling $P_{\lambda_1,\dots,\lambda_n}^{\gamma_1,\dots,\gamma_n}$. Finally define $P_{n+1}(\lambda_1, \dots, \lambda_{n+1}, \gamma_1, \dots, \gamma_{n+1}) = P_n(\lambda_1, \dots, \lambda_n, \gamma_1, \dots, \gamma_n) P_{\lambda_1,\dots,\lambda_n}^{\gamma_1,\dots,\gamma_n}(\lambda_{n+1}, \gamma_{n+1})$. P_{n+1} is the desired coupling.

Now that P_n has been constructed for all n you just let P be a subsequential limit of the P_n. P is a measure on $(\Lambda^2)^{\mathbf{N}}$, which may be identified with $(\Lambda^{\mathbf{N}})^2$.

302. Exercise. Show that P is the unique subsequential limit of the P_n and that it is a coupling of $(X_i)_{i=1}^{\infty}$ and $(Y_i)_{i=1}^{\infty}$ (that is, of the associated measures on $\Lambda^{\mathbf{N}}$).

303. Comment. You can couple double sided processes $(X_i)_{i=-\infty}^{\infty}$ and $(Y_i)_{i=-\infty}^{\infty}$ by induction as well; you just have to first choose a bijection $n : \mathbf{N} \to \mathbf{Z}$ and then couple the random variables in the order indicated by n. Note, however, that the language about subsequential limits needs to be modified a bit, albeit in a rather obvious way; alternatively, use couplings over "located words" (try to figure out what a located word is)[56] to induce a corresponding premeasure on cylinder sets of the product space and use Carathéodory's theorem to get to the measure on the product space. Details are left to the reader.

304. Example. A special case of Example 301: here we arrange that $P(X_{n+1}|X_1, X_2, \ldots, X_n)$ is independent of $(Y_i)_{i=1}^n$ and $P(Y_{n+1}|Y_1, Y_2, \ldots, Y_n)$ is independent of $(X_i)_{i=1}^n$.[57] This means, you simply couple the conditional probability of X_{n+1} given $(X_i)_{i=1}^n$ with the conditional probability of Y_{n+1} given $(Y_i)_{i=1}^n$.[58]

305. Example. A further special case, this time a greedy algorithm that attempts to provide a coupling that is supported on pairs of words that agree on a high number of coordinates. Let $(X_i)_{i=1}^{\infty}$ and $(Y_i)_{i=1}^{\infty}$ be processes on the alphabet $\{0, 1\}$. Start by coupling X_1 and Y_1 in such a way as to maximize the probability that the two coordinates are equal.[59] Now, having coupled $(X_i)_{i=1}^n$ with $(Y_i)_{i=1}^n$, couple the conditional probability of X_{N+1} given $(X_i)_{i=1}^n$ with the conditional probability of Y_{N+1} given $(Y_i)_{i=1}^n$ in a way that maximizes the probability that the $(n + 1)$st coordinates are equal.

---

[56] It's a function from a finite subset of $\mathbf{Z}$ to Λ.

[57] In other words, knowing the values of X_i, $1 \le i \le n$, no information about X_{n+1} is gained by further knowing the values of Y_i, $1 \le i \le n$. Notice that this does *not* say that knowing the values of Y_i, $1 \le i \le n$, won't give you information about the value of X_{n+1} in cases where you didn't previously know anything.

[58] In the notation of previous footnotes, $\mu_{\lambda_1,\ldots,\lambda_n}^{\gamma_1,\ldots,\gamma_n}(\lambda) = P(X_{n+1} = \lambda | X_i = \lambda_i, 1 \le i \le n)$, irrespective of the values of $\gamma_1, \ldots, \gamma_n$, etc.

[59] Recall that a coupling of X_1 and Y_1 is a measure α on $\{0, 1\} \times \{0, 1\}$ such that $\alpha\{(0, 0), (0, 1)\} = P(X_1 = 0)$ and $\alpha\{(0, 0), (1, 0)\} = P(X_2 = 0)$. We want to choose an α that maximizes $\alpha(\{(0, 0), (1, 1)\})$.

306. Example. *Gluing couplings together.* Suppose α_1 is a coupling of $(X_i)_{i=1}^n$ and $(Y_i)_{i=1}^n$ and α_2 is a coupling of $(Y_i)_{i=1}^n$ and $(Z_i)_{i=1}^n$. We want to give a coupling β of $(X_i)_{i=1}^n$, $(Y_i)_{i=1}^n$ and $(Z_i)_{i=1}^n$ such that restricting to the first two coordinates gives you back α_1 and restricting to the last two coordinates gives you back α_2. Here's how it works. First you put down an output of $(Y_i)_{i=1}^n$ in accordance with its probability law. Now you use α_1 to compute the conditional probability law μ_1 of $(X_i)_{i=1}^n$ given the output of $(Y_i)_{i=1}^n$ that has occurred. Next use α_2 to compute the conditional probability law μ_2 of $(Z_i)_{i=1}^n$ given the output of $(Y_i)_{i=1}^n$ that has occurred. Finally pick a coupling γ of μ_1 and μ_2. (You might use the independent coupling.) Notice that γ depends on the output of $(Y_i)_{i=1}^n$ you chose; you want to do this for all such outputs to get a measure on Λ^{3n} by integrating over all outputs.[60]

3.3. Applications of coupling

In this section we use coupling to prove a few theorems, including the Kolmogorov 0-1 law and a version of the renewal theorem.

307. Theorem. *(An application of coupling.) Let $(X_i)_{i=1}^\infty$ be a random walk (biased or unbiased). Let an even number $k \in \mathbf{N}$ and $\epsilon > 0$ be given. There exists N such that for every set B of integers, $|P(X_N \in B) - P(X_{N-k} \in B)| < \epsilon$.*

Sketch of proof. Let $Y_i = X_0$, $0 \le i \le k$, and let $(Y_i)_{i=k}^\infty$ be a random walk with the same parameter p as the (X_i) walk. Assume additionally that $(Y_i)_{i=k}^\infty$ is independent of (X_i) (otherwise, the assertion of the following exercise need not be true).

308. Exercise. There exists N such that $P(X_i = Y_i$ for some $k \le i \le N) > 1 - \epsilon$.   •

Couple $(X_i)_{i=1}^N$ and $(Y_i)_{i=1}^N$ inductively in the following way. Couple $(X_i)_{i=0}^k$ with $(Y_i)_{i=0}^k$ any way you like. Now, supposing $(X_i)_{i=0}^n$ and $(Y_i)_{i=0}^n$ have been coupled, for any pair of outputs $\lambda_0, \ldots, \lambda_n$ of (X_i) and $\gamma_0, \ldots, \gamma_n$ of (Y_i), couple the conditional probabilities of X_{n+1} given $(X_i)_{i=0}^n$ and of Y_{n+1} given $(Y_i)_{i=0}^n$ independently if $\lambda_n \ne \gamma_n$ and diagonally if $\lambda_n = \gamma_n$.[61] Denote by Q the resulting coupling of $(X_i)_{i=1}^N$ and $(Y_i)_{i=1}^N$.

---

[60] Some details: given words w, v, let $\mu_w(v) = \frac{\alpha_1(v,w)}{\alpha_1(\Lambda \times w)}$ and let $\nu_w(v) = \frac{\alpha_2(w,v)}{\alpha_2(w \times \Lambda)}$. Choose a coupling γ_w of μ_w and ν_w. Finally for words u, w, v put $c(u, w, v) = P(Y_1 \cdots Y_n = w)\gamma_w(u, v)$. c is the desired coupling of the three processes.

[61] The conditionals μ and ν of X_{n+1} and Y_{n+1}, respectively, are of course given by $\mu(\lambda_n + 1) = \nu(\gamma_n + 1) = p$, $\mu(\lambda_n - 1) = \nu(\gamma_n - 1) = 1 - p$. The coupling of μ and ν that is used for the marginal coupling is either $P(\lambda_n + 1, \gamma_n + 1) = p^2$, $P(\lambda_n - 1, \gamma_n + 1) = P(\lambda_n + 1, \gamma_n - 1) =$

309. Exercise. Show that $Q(\{(x, y) : x_N = y_N\}) > 1 - \epsilon$ and finish the proof.[62] □

310. Definition. If $(X_i)_{i=-\infty}^\infty$ is a process, the *n future* is the σ-algebra $\mathcal{F}_n$ generated by $(X_i)_{i=n}^\infty$. The *tailfield* $\mathcal{T}$ is the intersection, over all n, of the n future.[63]

311. Definition. Let $a, b \in \Lambda^{\mathbf{Z}}$. We write $a \sim b$ and say that a and b have the *same tail* if for some N, $a_n = b_n$ for every $n > N$.

312. Comment. Obviously $\sim$ is an equivalence relation.

313. Theorem. *Let* $(X_i)_{i=-\infty}^\infty$ *be a stationary process taking values in a countable alphabet* Λ. *Let* $A \subset \Lambda^{\mathbf{Z}}$ *be a measurable set. Then* $A \in \mathcal{T}$ *if and only if there is a set* B *such that* $\mu(A \triangle B) = 0$ *and* B *is a union of equivalence classes of* $\sim$.

Sketch of proof. For $n \in \mathbf{N}$ write $a \underset{n}{\sim} b$ if for every $i \geq n$, $a_i = b_i$. Then plainly $a \sim b$ if and only if there exists some n such that $a \underset{n}{\sim} b$. We make the following claim:

Claim. Let A be a measurable set. Then $A \in \mathcal{F}_n$ if and only if there is some B such that $\mu(A \triangle B) = 0$ and B is a union of equivalence classes of $\underset{n}{\sim}$.

To prove the claim, start by letting $(B_i)_{i=1}^\infty$ be the algebra generated by all sets of the form $X_i^{-1}(\lambda)$, where $\lambda \in \Lambda$ and $i \geq n$. This algebra is dense in $\mathcal{F}_n$. Carry out the construction of the canonical factor $(\Omega', \mathcal{A}', \mu')$ per Example 56 using the sequence (B_i) and observe that the fibers are precisely the equivalence classes under $\underset{n}{\sim}$.

314. Exercise. Use this observation to prove the claim. *Hint: for one direction, let* $A \in \mathcal{F}_n$ *and take* $B = \{x : E(1_A|\mathcal{B}) = 1\}$. *Use Exercise 117 (b). Conversely, let* A *be measurable and a union of equivalence classes, i.e. fibers. Observe that* $A = \pi^{-1}(B)$ *for some* $B \subset \Omega'$. *Apply Exercises 49 and 61.*

315. Exercise. Now use the claim to finish off the proof.[64] □

$p(1 - p)$, $P(\lambda_n - 1, \gamma_n - 1) = (1 - p)^2$ in the event $\lambda_n \neq \gamma_n$ and $P(\lambda_n + 1, \lambda_n + 1) = p$, $P(\lambda_{n-1}, \lambda_n - 1) = 1 - p$ in the event $\lambda_n = \gamma_n$.

[62] It is useful to notice that $P(X_N \in B) = Q(\{(x, y) : x_N \in B\})$ and $P(X_{N-k} \in B) = Q(\{(x, y) : y_N \in B\})$.

[63] That is, $\mathcal{T} = \bigcap_{n=1}^\infty \mathcal{F}_n$. Obviously it depends on X.

[64] Some details: suppose $A \in \mathcal{T}$. Then $A \in \mathcal{F}_n$ for every n, so there exist sets (B_n) such that $\mu(A \triangle B_n) = 0$ for every n and B_n is a union of equivalence classes under $\underset{n}{\sim}$. Let $B = \bigcup_{j=1}^\infty \bigcap_{i=j}^\infty B_i$. Plainly $\mu(A \triangle B) = 0$. Suppose $x \in B$ and $x \sim y$. Then there exists N_1 such that $x \underset{n}{\sim} y$ for all $n \geq N_1$. Now choose N_2 such that $x \in B_n$ for all $n \geq N_2$. Then for every

316. Definition. A process is said to have *trivial tail* if every set in the tailfield has measure 0 or 1.

317. Theorem. *(Kolmogorov 0-1 law; see e.g. Dudley 2002, Theorem 8.4.4.) Any independent stationary process has trivial tail.*

Sketch of proof. We take $(X_i)_{i=1}^{\infty}$ to be our independent process, which we are assuming one sided for convenience. We assume the X_i are random variables into a countable alphabet Λ and form the associated space $(\Omega, \mathcal{A}, \mu)$, where $\Omega = \Lambda^{\mathbb{N}}$, etc. As is our standard convention, we view the random variables X_i as being defined on Ω. Let B be a member of the tailfield; by adding and deleting sets of measure zero we can assume that B is a union of equivalence classes under $\sim$.

Take a cylinder set C that closely approximates B and define a measure ν by $\nu(A) = \frac{\mu(A \cap C)}{\mu(C)} = P_\mu(A|C)$, $A \in \mathcal{A}$. The same sequence $(X_i)_{i=1}^{\infty}$, only defined on the measure space $(\Omega, \mathcal{A}, \nu)$, is called the *conditioned process* (conditioned on C, that is). Now we want to couple the measures μ and ν. That is, we want to couple the original process and the conditioned process. We do it inductively. Couple any way you like until, eventually, you get beyond all the X_i for which i is one of the coordinates involved in the definition of C. For subsequent n, you have

$$P_\mu(X_n = \lambda|X_1, \ldots, X_{n-1}) = P_\nu(X_n = \lambda|X_1, \ldots, X_{n-1}) = c(\lambda)$$

for every letter λ and every valuation of $X_1, X_2, \ldots, X_{n-1}$. For the induction step, couple these (equal) marginals together with the diagonal coupling. Let P be the resulting coupling.

318. Exercise. Show that P is supported on pairs of words (x, y) with $x \sim y$ and use this to show that $\mu(B) = \nu(B)$. *Hint: show that both are equal to $P(B \times B)$.* Conclude that $\mu(B) \in \{0, 1\}$. □

319. Digression: the renewal theorem. For our final application in this section, we present a proof via couplings of one of the main theorems on Markov chains, the renewal theorem. We feel that this proof may be of some interest even to those already well acquainted with the theorem, in that it is arguably more intuitive and natural than are standard proofs. We won't use this material later, so disinterested readers are advised to skip. First, a bit of background.

$n \geq \max\{N_1, N_2\}$, $x \in B_n$ and $x \underset{n}{\sim} y$, hence $y \in B_n$. This implies $y \in B$ and B is a union of equivalence classes of $\sim$. Conversely, suppose A is measurable and a union of equivalence classes under $\sim$. Let $n \in \mathbb{N}$, let $x \in A$ and suppose $x \underset{n}{\sim} y$. Then $x \sim y$, so $y \in A$. Hence A is a union of equivalence classes under $\underset{n}{\sim}$. By the claim, $A \in \mathcal{F}_n$ and since n is arbitrary, $A \in \mathcal{T}$.

320. Definition. Let Λ be a finite alphabet. (Without loss of generality, we may assume that $\Lambda = \{1, 2, \ldots, k\}$.) A *possibly non-stationary Markov chain* (or *pnm*) on Λ is a stochastic process $(X_n)_{n=1}^{\infty}$ satisfying the following:

(1) for every $n \in \mathbf{N}$, X_n takes values in Λ, and
(2) for some constants $p(i, j)$, $1 \leq i, j \leq k$, for all $n \in \mathbf{N}$ one has

$$P\left(X_n = i_n | X_t = i_t, 1 \leq t \leq n - 1\right) = P\left(X_n = i_n | X_{n-1} = i_{n-1}\right)$$
$$= p(i_{n-1}, i_n).$$

(Note in particular that this conditional probability does not depend on X_i, $i < n - 1$, nor does it depend on n.)

321. Definition. The constants $p(i, j)$ are called *transition probabilities*.

322. Definition. Let $i, j \in \Lambda$. We say that there is a *path of length* $n \in \mathbf{N}$ from i to j if there exist $c_0, c1, c2, \ldots, c_n \in \Lambda$ such that:

(1) $i = c_0$,
(2) $j = c_n$, and
(3) $p(c_{t-1}, c_t) > 0$, $1 \leq t \leq n$.

323. Exercise. Show that there is a path of length n from i to j if and only if for some t, $P(X_{t+n} = j | X_t = i) > 0$.

324. Definition. If there is a path from i to j and a path from j to i, then we say that i and j *communicate*. If i fails to communicate with itself, we say that i is *strongly transient*.

325. Definition. If i is not strongly transient, let

$$\theta(i) = \{n \in \mathbf{N} : \text{there is a path of length } n \text{ from } i \text{ to itself}\},$$

and let $g(i)$ be the greatest common divisor of $\theta(i)$.

326. Theorem. *If i is not transient, there exists n such that $\theta(i)$ contains every multiple of $g(i)$ greater than n.*

Idea of proof. First show that $\theta(i)$ is a sub-semigroup of $(\mathbf{N}, +)$ (i.e. a non-empty subset of $\mathbf{N}$ that is closed under sums). Next show that every sub-semigroup G of $\mathbf{N}$ contains every sufficiently large multiple of its greatest common divisor g. Do this as follows. Let S be a finite subset of G such that g is a linear combination of the members of S with integer coefficients. Next let t be the sum of the members of S, and show that for all sufficiently large k, $kt + ig$, $0 \leq i < t$, can be written as a linear combination of the members of S with non-negative integer coefficients. $\square$

327. Theorem. *(The renewal theorem.) Suppose $(X_i)_{i=0}^\infty$ is a pnm and let $\epsilon >$ 0. Suppose $a, b, c \in \Lambda$ communicate with one another. Suppose further that $g(a) = 1$ and that any state to which a has a path communicates with a. Then there exists N such that whenever $n, m > N$, $\left| P(X_m = c | X_0 = a) - P(X_n = c | X_0 = b) \right| < \epsilon$.*

The proof of this theorem is going to strongly resemble the proof of Theorem 307. Before giving ideas of the proof, we have the following:

328. Lemma. *Let $d, e,$ and f be any states in Λ communicating with a. Suppose two processes $(Y_i)_{i=1}^\infty$ and $(Z_i)_{i=1}^\infty$ are independent of one another and have the exact same alphabet and transition probabilities as (X_i), but start from $Y_0 = d$ and $Z_0 = e$, respectively. Then for any $\delta > 0$ there exists M such that $P(Y_i = f = Z_i \text{ for some } 1 \le i \le M) > 1 - \delta$.*

Sketch of a proof of the lemma.

329. Exercise. Let S be the set of states communicating with a. Show that for all $s \in S$ and all large enough n, $P(X_n = f | X_0 = s) > 0$. *Hint: use Theorem 326.*

330. Exercise. Use the foregoing exercise to show that there exist J and some $d > 0$ such that for all $s \in S$ and all $k > J$, $P(Y_{J+k} = f | Y_k = s) = P(Z_{J+k} = f | Z_k = s) > d$. (Note one must have for this that $P(Y_k = s) > 0$ and $P(Z_k = s) > 0$; use the previous exercise.) Conclude by independence that for any $s_1, s_2 \in S$, $P(Y_{J+k} = f = Z_{J+k} | Y_k = s_1, Z_k = s_2) > d^2$.

331. Exercise. Show that the proof of the lemma can be finished off by choosing t so large that $(1 - d^2)^t < \delta$ and letting $M = Jt$.   □

Idea of proof. (Of Theorem 327.) With the lemma in hand, we may proceed exactly as in the proof of Theorem 307. The details are left largely to the reader, but we offer presently a sort of "storybook" interpretation of what's going on. Let $N = M$. If $n \ge m > N$, choose independent (Y_i) and (Z_i) processes as in the lemma, but with $Y_0 = a$ and $Z_0 = Z_1 = \cdots = Z_{n-m} = b$. Think of the (Y_i) process as a criminal running from a detective (represented by the (Z_i) process). The detective gives the criminal a head start of $n - m$ time steps. Now, $n - m$ might be very large, much larger than, say, m. Even so, the detective is still very likely to catch the criminal within m time steps. This is where the proof differs from that of Theorem 307; here, the criminal can run until the cows come home and he still can't get too far away from the detective because he's running on a finite set. So anyway you couple the two processes together in such a way to reflect that if the detective ever catches the criminal he handcuffs him and they remain together thereafter. Since by time

m they are together with probability at least $1 - \epsilon$, it is certainly the case that $|P(Y_n = c) - P(Z_n = c)| < \epsilon$. But $P(Y_n = c) = P(X_n = c | X_0 = a)$ and $P(Z_n = c) = P(X_m = c | X_0 = b)$. $\qquad\qquad\qquad\qquad\qquad\qquad\square$

3.4. The dbar and variation distances

We use coupling to define two metrics on the family of stationary processes on an alphabet Λ. This material is essential and should not be skipped.

332. Definition. The *mean Hamming distance* between two words of the same length is the fraction of the located letters that are different.[65] For example, $d(\text{stashed, plaster}) = \frac{4}{7}$. (Notice that the ts don't count as being the same; they don't occur in the same place.)

333. Exercise. Prove that, on the space of words of a given length n, the mean Hamming distance is a metric.

334. Definition. Let Λ be an alphabet and $n \in \mathbf{N}$. The *dbar distance* $\overline{d}(\mu, \nu)$ between two measures on the words of length n from Λ is the infimum of the expected mean Hamming distance between coupled words, taken over all couplings of the measures.[66]

335. Exercise. Show that the infimum of the previous definition is achieved; that is, show it is in fact a minimum.

336. Exercise. Let $f : \mathbf{N} \to [0, \infty)$ be a function satisfying $f(0) = 0$, $f(n - 1) \leq f(n) \leq f(n - 1) + 1$ and $f(kn) \leq kf(n)$ for all $k, n \in \mathbf{N}$. Show that $\lim_{n \to \infty} \frac{f(n)}{n}$ exists.

337. Comment. The dbar distance is closely related to various distance metrics that have been around for a long time, going by the names Vasershtein distance, Monge–Kantorovich transportation metric, earth mover's distance, etc. See the two volume Rachev and Rüschendorf (1998) for a history.

Now let $(X_i)_{i=1}^{\infty}$ and $(Y_i)_{i=1}^{\infty}$ be stationary processes on a countable alphabet Λ. As usual (see the discussion in 152) we assume that each of the random variables in these sequences has $\Omega = \Lambda^{\mathbf{N}}$ as the domain space.

---

[65] For $w = \lambda_1 \lambda_2 \cdots \lambda_n$ and $\nu = \gamma_1 \gamma_2 \cdots \gamma_n$, one has $d(w, \nu) = \frac{\#\{i : \lambda_i \neq \gamma_i\}}{n}$.

[66] Let μ and ν be measures on Λ^n. Recall that a coupling of μ and ν is a measure P on $\Lambda^n \times \Lambda^n$, that is, a measure on pairs of words (w, ν). The expected mean Hamming distance on pairs relative to P is $\sum_{w,\nu} P(w, \nu) d(w, \nu)$. $\overline{d}(\mu, \nu)$ is the infimum of this expectation over all couplings P.

Recall (again see 152) that the processes $(X_i)_{i=1}^{\infty}$ and $(Y_i)_{i=1}^{\infty}$ are characterized by the measures μ and ν they induce on Ω. Let $\mu_n(\lambda_1 \cdots \lambda_n) = P(X_1 = \lambda_1, \ldots, X_n = \lambda_n)$ and $\nu_n(\lambda_1 \cdots \lambda_n) = P(Y_1 = \lambda_1, \ldots, Y_n = \lambda_n)$. μ_n and ν_n are measures on Λ^n, and we denote their dbar distance by $g(n)$.

338. Exercise. Show that $g = \lim_{n \to \infty} g(n)$ exists. *Hint: show that $f(n) = ng(n)$ satisfies the conditions of Exercise 336.*

339. Definition. The limit g of the previous exercise is called the *dbar distance* between the processes $(X_i)_{i=1}^{\infty}$ and $(Y_i)_{i=1}^{\infty}$, and is denoted by $\overline{d}((X_i), (Y_i))$.

340. Exercise. The dbar distance $\overline{d}((X_i), (Y_i))$ can always be achieved by a stationary coupling.[67] *Hint: treat Λ^2 as a countable alphabet and view couplings between two measures on words of length n as measures on length n words over Λ^2. Use the extended monkey method on an "infimal" sequence of couplings.*

341. Definition. Let $(\Omega, \mathcal{A}, \mu)$ be a probability space and let $n \in \mathbf{N}$. By an *n-element, measurable ordered partition of Ω* (generally, we shall say simply "ordered partition of Ω") we mean an n-tuple $(A_1, \ldots, A_n)$ of sets in $\mathcal{A}$ such that $\Omega = \bigcup_{i=1}^{n} A_i$ and $A_i \cap A_j = \emptyset$ for $1 \le i \ne j \le n$.

342. Definition. Let $(\Omega, \mathcal{A}, \mu)$ and $(X, \mathcal{B}, \nu)$ be probability spaces having ordered partitions $A = (A_1, \ldots, A_n)$ and $B = (B_1, \ldots, B_n)$, respectively. The *variation distance* between A and B is

$$v(A, B) = \inf\left\{1 - \alpha\left(\bigcup_{i=1}^{n} A_i \times B_i\right) : \alpha \text{ is a coupling of } \mu, \nu\right\}$$

$$= 1 - \sup\left\{\alpha\left(\bigcup_{i=1}^{n} A_i \times B_i\right) : \alpha \text{ is a coupling of } \mu, \nu\right\}.$$

343. Important comment. To be technically precise we really ought, in the above definition, to say the variation distance between the pair (A, μ) and the pair (B, ν). However, we shall usually omit reference to the underlying measures when they are understood.

344. Exercise. Show that the infimum of Definition 342 is in fact a minimum.

---

[67] What this means is that there is a measure c on $(\Lambda^{\mathbf{N}})^2$ (which may be identified with $(\Lambda^2)^{\mathbf{N}}$) such that for the induced measures c_n on $(\Lambda^2)^n$, which are defined by

$$c_n\binom{\lambda_1 \cdots \lambda_n}{\gamma_1 \cdots \gamma_n} = c\left(\{((x_i)_{i=1}^{\infty}, (y_i)_{i=1}^{\infty}) : x_i = \lambda_i, y_i = \gamma_i, 1 \le i \le n\}\right),$$

one has $\lim_{n \to \infty} \sum_{w, v \in \Lambda^n} c_n(w, v) d(w, v) = \overline{d}((X_i), (Y_i))$.

345. Definition. Let p and q be two probability measures on a finite set $\{x_1, \ldots, x_n\}$. The *variation distance* $v(p, q)$ between p and q is $\frac{1}{2} \sum_{i=1}^{n} |p(x_i) - q(x_i)|$.

346. Exercise. Show that the variation distance between p and q is exactly the same as the variation distance between the ordered partition $(\{1\}, \ldots, \{n\})$ (under the measure p) and the same ordered partition $(\{1\}, \ldots, \{n\})$ (under the measure q).

347. Exercise. Show that when $\mu = v$, $v(A, B) = \frac{1}{2} \sum_{i=1}^{n} \mu(A_i \triangle B_i)$.

348. Exercise. Generalize the foregoing definition and exercise to measures on a countable set $\Omega = \{x_1, x_2, \ldots\}$.

349. Comment. An important special case of the above is when Ω is the set of words over an alphabet Λ of a given finite length. Note: saying the variation distance between two measures on Ω is small is much stronger than saying the dbar distance between them is small.

350. Definition. Let X_1 and X_2 be real-valued random variables. Define measures μ_i on $\mathbf{R}$, $i = 1, 2$, by the rule $\mu_i(B) = P(X_i \in B)$. A *coupling of X_1 and X_2* is a coupling of μ_1 and μ_2. The *variation distance between X_1 and X_2* is the minimum of the probability that $x \neq y$ over all couplings of X_1 and X_2.[68]

3.5. Preparation for the Shannon–Macmillan–Breiman theorem

As a final application of coupling, we give a lemma designed to prepare for the proof of the most important theorem of the next chapter, the Shannon–McMillan–Breiman theorem.

351. Notation. Let $(a_i)_{i=1}^{\infty}$ be a sequence of real numbers. We write $C \lim_i a_i$ as shorthand for $\lim_{n \to \infty} \frac{1}{n} \sum_{i=1}^{n} a_i$. $C \lim \sup$ and $C \lim \inf$ are defined similarly.

352. Definition. For $N > 0$, let f_N be the function defined as follows: $f_N(x) = x$, if $|x| > N$, and $f_N(x) = 0$ otherwise.

353. Definition. A sequence $(a_i)_{i=1}^{\infty}$ is *essentially bounded* if

$$\lim_{N \to \infty} C \lim \sup_i |f_N(a_i)| = 0.$$

---

[68] To be more precise, the variation distance between X and Y is the infimum, over all couplings α of μ_1 and μ_2, of $\alpha(\{(x, y) : x \neq y\})$. One can show that this is a minimum.

354. Comment. The "C" in the notation "$C\lim$" stands, of course, for "Cesàro". Essentially bounded sequences have the property that the "unbounded" part does not interfere with Cesàro summation of the sequence, allowing one to assume without loss of generality in many situations that one is actually dealing with a bounded sequence.

355. Theorem. *Let $(a_i)_{i=1}^{\infty}$ be any sequence of non-negative reals such that $\sum_{i=1}^{\infty} ia_i < \infty$. If (X_i) is a process, where the X_i are real-valued random variables, satisfying*

$$P(i \le |X_n| < i + 1|(X_i)_{i=1}^{n-1})(x) < a_i$$

for all i, all n and a.e. x, then for a.e. x the sequence $\left(X_i(x)\right)_{i=1}^{\infty}$ is essentially bounded.

356. Comment. The hypotheses of this theorem should strike the reader as outrageously strong; hence the reasonableness of such a strong conclusion.

Idea of proof. Choose N large enough that $a_{N-1} + a_N + a_{N+1} + \cdots < 1$. Pick a random variable Y_N with distribution $P(Y_N = 0) = 1 - (a_{N-1} + a_N + a_{N+1} + \cdots)$, $P(Y_N = i) = a_{i-1}, i > N$. Let $(Y_N^{(i)})_{i=1}^{\infty}$ be an independent sequence of random variables each with the distribution of Y_N. Now couple $(|F_N(X_i)|)_{i=1}^{\infty}$ to $(Y_N^{(i)})_{i=1}^{\infty}$ inductively so that $|f_N(X_i)| < Y_N^{(i)}$ for all i. (Here we make our coupling so that $i - 1 < f_N(X_i) < i$ whenever $Y_n = i, i > N$.) Now check that when N is large, $\frac{1}{n}\sum_{i=1}^{n}(Y_N^{(i)})$ approaches something small.

Sketch of proof. Let $\epsilon > 0$ be arbitrary. We must show that for a.e. x there is some N_0 such that for all $N > N_0$, one has $\limsup_{n\to\infty} \frac{1}{n}\sum_{i=1}^{n}|f_N(X_i)| < \epsilon$. We will do even more: we will show that N_0 need not depend on x.

For all $N \in \mathbf{N}$ large enough that $\sum_{i=N-1}^{\infty} a_i < 1$, let Y_N be a random variable such that $P(Y_N = 0) = 1 - \sum_{i=N-1}^{\infty} a_i$ and for $i \ge N$, $P(Y_N = i) = a_{i-1}$.

357. Exercise. Use the summability hypothesis of the theorem to show that for some N_0, $E(Y_N) < \epsilon$ for all $N > N_0$. •

Now suppose $N > N_0$. Let $(Y_{N,i})_{i=1}^{\infty}$ be a sequence of random variables that is independent, independent of the sequence (X_i), and such that the distribution of $Y_{N,i}$ is the same as the distribution of Y_N.

358. Exercise. Inductively couple $\left(|F_N(X_i)|\right)_{i=1}^{\infty}$ with $(Y_{N,i})_{i=1}^{\infty}$ in such a way that $|f_N(X_i)| < Y_{N,i}$ a.e. for all i. *Hint: use the conditional probability* *hypothesis of the theorem, together with our independence assumptions, to guarantee that $Y_{N,i} = j$ whenever $j - 1 \leq |f_N(X_i)| < j$.*   •

The proof is now completed by an application of Theorem 210. To wit:

$$\limsup_{n \to \infty} \frac{1}{n} \sum_{i=1}^{n} |f_N(X_i)| \leq \lim_{n \to \infty} \frac{1}{n} \sum_{i=1}^{n} Y_{N,i} = E(Y_N) < \epsilon$$

a.e.   ☐

4

Entropy

The notion of entropy in ergodic theory has its roots in C. Shannon's (1948) classic paper, which launched information theory as a well-developed science. The *entropy* of an ergodic system $(\Omega, \mathcal{A}, \mu, T)$ is a non-negative number that in a certain sense measures the amount of randomness in the system. In this chapter we'll define entropy in a non-standard way and show that our definition is equivalent to various other classical formulations. The reader should note that there are problems with extending the notion of entropy to non-ergodic systems; in particular, the many equivalent formulations turn out not to be so when one doesn't assume ergodicity.

4.1. The 3-shift is not a factor of the 2-shift

The theorem we prove in this section as an introduction to entropy was open for a long time, even though a lot of very good mathematicians tried to do it. However, we use essentially nothing in our short proof of it. This is the benefit of hindsight in mathematics!

359. Comment. We are about to define the stationary process known as the n-shift. One should think of it as being obtained by taking a fair n-sided die and tossing it doubly infinitely many times.

Let $n \in \mathbf{N}$ and let $\Lambda_n = \{1, 2, \ldots, n\}$, considered as a finite alphabet having n letters. Put $\Omega_n = \Lambda_n^{\mathbf{Z}}$ and let $T_n : \Omega_n \to \Omega_n$ be the shift, $T_n \gamma(i) = \gamma(i+1)$. For a cylinder set $C = \{x \in \Omega_n : x_{i_1} = a_1, \ldots, x_{i_t} = a_t\}$, put $p(C) = n^{-t}$. p extends additively to the algebra generated by cylinder sets and to a T_n-invariant measure μ on a σ-algebra $\mathcal{A}_n$ by Carathéodory's theorem.

360. Definition. The measure-preserving system $(\Omega_n, \mathcal{A}_n, \mu_n, T_n)$ described above is called the *n-shift*.

361. Comment. Recall Definition 222: if B is a set of words of length n we denote by $\varphi(B)$ the set of doubly infinite words $(a_i)_{i=-\infty}^{\infty}$ such that $a_1 a_2 \cdots a_n \in B$.

362. Definition. Let W be a set of words of length n. The number of words in W is of course $|W|$. The *exponential number of words* in W is $H = \frac{\log |W|}{n}$.

(Equivalently, $|W| = 2^{Hn}$.) The *size* of W is $\mu(\varphi(W))$. The *exponential size* of W is $\dfrac{-\log\mu(\varphi(W))}{n}$.

363. Comment. Sometimes we may say "a small exponential number of words of length n"; by this we mean a set of words W with, say $\dfrac{\log|W|}{n} = \epsilon$, where ϵ is small.[69] For example, if $|W| = 2^{.001n}$ then we would probably say that about W. We can speak of the size or exponential size of a single word w; by this we just mean the size or exponential size of the singleton $\{w\}$. Notice that the exponential size of a word w_1 of length n is *greater* than the exponential size of a word w_2 of length n when the size of w_1 is *less* than the size of w_2.

In the following exercise you are asked to prove a technical fact that will aid in the proof of the next theorem.

364. Exercise. Let w be a word of length n on an alphabet having three letters. Show that the number of words u such that $\overline{d}(u, w) < \epsilon$ is exponentially small (when ϵ is small).[70] *Hint: recall Stirling's formula, $n! \approx n^n e^{-n}\sqrt{2\pi n}$.*

365. Theorem. *The 3-shift is not a factor of the 2-shift. That is, $(\Omega_3, \mathcal{A}_3, \mu_3, T_3)$ is not a factor of $(\Omega_2, \mathcal{A}_2, \mu_2, T_2)$.*

Idea of proof. Suppose $\pi : \Omega_2 \to \Omega_3$ is a homomorphism. Approximate the inverse image of the canonical three-set partition on Ω_3 with a three-set partition of cylinder sets in Ω_2 having maximum length, say, 9. Presume for the moment, that this approximation is actually exact. Then each subword of length 200 of a point $x \in \Omega_3$ is determined by a distinct word of length 208 in $\pi^{-1}(x)$. This is impossible because $2^{208} < 3^{200}$. However, since the cylinder set partition is only an approximation, words of length 208 determine words of length 200 only after some small proportion of letters are altered (on most of the space). Even so, by Exercise 364 each word of length 208 can still account for only an exponentially small number of words of length 200, which isn't enough.

Sketch of proof. Suppose $\pi : \Omega_2 \to \Omega_3$ is a homomorphism. Let $A_i = \{x \in \Omega_3 : x_0 = i\}$, $i = 1, 2, 3$. Put $\delta = \frac{1}{1000}$ and choose ϵ as in the footnote cited in the last exercise, that is such that for all sufficiently large n, $\binom{n}{\lceil \epsilon n \rceil}3^{\lceil \epsilon n \rceil} < 2^{\delta n}$.

---

[69] Technically, what one usually has in mind is a sequence (W_n) of sets of words of length n, with $\frac{\log|W_n|}{n} \to 0$; cf. the definition of "exponentially fat" below.

[70] The number of such words is at most $\binom{n}{\lceil \epsilon n \rceil}3^{\lceil \epsilon n \rceil}$. What you must show here is that for every $\delta > 0$, there exists $\epsilon > 0$ such that for all sufficiently large n, $\binom{n}{\lceil \epsilon n \rceil}3^{\lceil \epsilon n \rceil} < 2^{\delta n}$.

366. Exercise. There exist sets $C_1, C_2, C_3 \subset \Omega_2$ satisfying the following:

(a) each C_i is a finite union of cylinder sets;
(b) the C_i are pairwise disjoint;
(c) $\Omega_2 = C_1 \cup C_2 \cup C_3$;
(d) $\mu_2\big(C_i \triangle \pi^{-1}(A_i)\big) < \frac{\epsilon}{6}, i = 1, 2, 3$. •

Choose k large enough that each C_i is supported on $\{-k, -k+1, \ldots, k\}$. Define $\gamma : \Omega_2 \to \{1, 2, 3\}$ by $\gamma(x) = i$ if and only if $x \in C_i$ and set $B = \{x \in \Omega_2 : \gamma(x) \neq \pi(x)_0\}$.

367. Exercise. Show that $\mu_2(B) < \frac{\epsilon}{2}$, whence $P(\sum_{i=0}^{n-1} T_2^{-i} 1_B > n\epsilon) < \frac{1}{2}$ for every n. •

Fix a large n and define $\alpha, \beta : \Omega_2 \to \{1, 2, 3\}^{\{0, 1, \ldots, n-1\}}$ by $\alpha(x) = \gamma(x)\gamma(Tx)\gamma(T^2 x) \cdots \gamma(T^{n-1}x)$ and $\beta(x) = \pi(x)_0 \pi(x)_1 \cdots \pi(x)_{n-1}$.

368. Exercise. Show that $P\big(d(\alpha(x), \beta(x)) \leq \epsilon\big) > \frac{1}{2}$, and that $|\text{range } \alpha| \leq 2^{n+2k}$. •

Let $D = \big\{z \in \{1, 2, 3\}^{\{0, 1, \ldots, n-1\}} : \bar{d}(z, y) \leq \epsilon \text{ for some } y \in \text{range } \alpha\big\}$. Then $|D| \leq 2^{n+2k} \binom{n}{\lceil \epsilon n \rceil} 3^{\lceil \epsilon n \rceil} < \frac{1}{3} 3^n$. This implies that $\mu_3\big(\varphi(D)\big) < \frac{1}{2}$. On the other hand, setting $E = \big\{x \in \Omega_2 : \bar{d}(\alpha(x), \beta(x)) \leq \epsilon\big\}$, we have $\mu_2(E) > \frac{1}{2}$. But $\pi(E) \subset \varphi(D)$. □

4.2. The Shannon–McMillan–Breiman theorem

In this section we prove the most basic major theorem of entropy theory. We use a fact we call the *square lemma*. We feel that this lemma is important in that the technique it employs is a universal one in analysis.

369. Exercise. Let $(a_i)_{i=1}^\infty$ be a real-valued sequence. Let $\epsilon, \delta > 0$ and $M > 0$. If $|a_i| \leq M$ for all i and $|a_i| \leq \epsilon$ for all $i \in E^c$, where $\limsup_{n \to \infty} \frac{|E \cap \{1,2,\ldots,n\}|}{n} \leq \delta$, then $C \limsup a_i \leq \epsilon + \delta M$.

370. Exercise. Let (Y_i) be a real-valued, non-negative stationary process and suppose that for some $\epsilon > 0$, $E(Y_1) \leq \epsilon^2$. Then $P(C \lim Y_i \geq \epsilon) \leq \epsilon$.

371. Square lemma. *Let* $(X_{i,j})_{i,j=1}^\infty$ *be real-valued random variables such that the columns form a stationary process; that is, such that for each fixed k, the sequence of vector-valued random variables* $\big((X_{1,j}, X_{2,j}, \ldots, X_{k,j})\big)_{j=1}^\infty$ *forms a stationary process.[71] Now suppose that* $(X_j)_{j=1}^\infty$ *is a sequence of*

---

[71] So, letting $\mathbf{X}_j = (X_{1,j}, X_{2,j}, \ldots, X_{k,j})$, for measurable sets $\mathbf{A}_t \in \mathbf{Z}^k$, $1 \leq t \leq s$, one has $P(\mathbf{X}_{j_1+t} \in \mathbf{A}_t, 1 \leq t \leq s) = P(\mathbf{X}_{j_2+t} \in \mathbf{A}_t, 1 \leq t \leq s)$.

random variables such that $\lim_{i\to\infty} X_{i,j} = X_j$ *a.e. for every* j. *If both of the sequences* $(X_j)_{j=1}^{\infty}$ *and* $(X_{i,i})_{i=1}^{\infty}$ *are essentially bounded a.e., then*

$$\lim_{n\to\infty} \frac{1}{n} \sum_{i=1}^{n} X_{i,i} = \lim_{n\to\infty} \frac{1}{n} \sum_{i=1}^{n} X_i$$

a.e. (In particular, both these limits exist a.e.)

Idea of proof. First use essential boundedness to reduce to the case where the $(X_{i,j})$ are uniformly bounded. Then show that $(X_j)_{j=1}^{\infty}$ is stationary, so that by Birkhoff, $\frac{1}{n}\sum_{i=1}^{n} X_i$ converges. Now it's enough to show that $C \lim_i |X_{i,i} - X_i| = 0$ a.e. Fix ϵ and N. Let $Y_j = 1$ if there are some $i, k > N$ such that $|X_{i,j} - X_{k,j}| > \epsilon$, $Y_j = 0$ otherwise. Then for large N, $\frac{1}{n}\sum_{i=1}^{n} Y_i$ converges a.e. to something small.

Sketch of proof. It is an easy exercise to reduce the general case to the case where all of the $X_{i,j}$ are uniformly bounded.[72] (The impetus for the essential boundedness concept is precisely so we can make this assumption.) Indeed we may assume without loss of generality that $|X_{i,j}| \le 1$ a.e. Next one shows that the sequence $(X_j)_{j=1}^{\infty}$ is stationary, which we leave to the reader.[73] Now it follows by Birkhoff that $\lim_{n\to\infty} \frac{1}{n}\sum_{i=1}^{n} X_i$ exists a.e. Therefore, it will be sufficient to show that for arbitrary $\epsilon > 0$,

$$P(C \limsup |X_{i,i} - X_i| \ge 2\epsilon) \le \epsilon.$$

Fix $\epsilon > 0$. Choose N so large that $P\big(\exists i, k > N \text{ such that } |X_{i,1} - X_{k,1}| > \epsilon\big) < \epsilon^2$. Now for each j let $Y_j = 1$ if there are $i, k > N$ such that $|X_{i,j} - X_{k,j}| > \epsilon$ and $Y_j = 0$ otherwise. Plainly $E(Y_1) < \epsilon^2$ and if $i > N$ and $Y_i(x) = 0$ then $|X_{i,i}(x) - X_i(x)| \le \epsilon$.

372. Exercise. Show that $(Y_i)_{i=1}^{\infty}$ is stationary. *Hint: let* $Y_{r,j} = 1$ *if* $|X_{i,j} - X_{k,j}| > \epsilon$ *for some* $N < i, k < r$. *Then for fixed* r $(Y_{r,j})_{j=1}^{\infty}$ *is stationary and* $\lim_r Y_{r,j} = Y_j$ *a.e.* ●

Now, for a.e. x, $E_x = \{n : |X_{i,i}(x) - X_i(x)| > \epsilon\} \subset \{n : Y_n(x) = 1\}$, and hence by Exercise 370, $\limsup_{n\to\infty} \frac{|E_x \cap \{1,...,n\}|}{n} \le \epsilon$ with probability $\ge$

---

[72] Here are some details. We'll assume $X_{i,j}$ is non-negative here, leaving the general case to the reader. Notice that $X_{i,j} = M_{i,j} + f_M(X_{i,j})$; here $M_{i,j} = \max\{X_{i,j}, M\}$. Moreover $\lim_{i\to\infty} f_M(X_{i,j}) = f_M(X_j)$, which implies that $X_j = M_j + f_M(X_j)$, where $M_j = \lim_i M_{i,j}$. So one need only choose M so large that $C \limsup_i |f_M(X_{i,i})|$ and $C \limsup |f_M(X_i)|$ are small.

[73] The basic idea here is that, e.g. $P(X_1 \in A_1, X_2 \in A_2) \approx P(X_{n,1} \in A_1, X_{n,2} \in A_2) = P(X_{n,k} \in A_1, X_{n,k+1} \in A_2) \approx P(X_k \in A_1, X_{k+1} \in A_2)$ when n is large.

$1 - \epsilon$. Meanwhile for these values of x, $C \limsup |X_{i,i}(x) - X_i(x)| \leq 2\epsilon$ by Exercise 369.   □

373. Theorem. *(Shannon–McMillan–Breiman theorem; Shannon 1948, McMillan 1953, Breiman 1957.) Let $(Y_i)_{i=-\infty}^{\infty}$ be a stationary process on a finite alphabet Λ. For a.e. $x \in \Lambda^{\mathbf{Z}}$,*

$$H(x) = \lim_{n \to \infty} -\frac{1}{n} \log P(Y_t = x_t, 1 \leq t \leq n)$$

exists.[74] *If the process is ergodic then the limit is constant a.e.*

Idea of proof. The idea is to use the square lemma to show that

$$-\frac{1}{n} \log \left(P(b_1 b_2 \cdots b_n) \right)$$

$$= -\frac{1}{n} \log \left(P(b_1) \right) - \frac{1}{n} \log \left(P(b_2|b_1) \right) - \cdots - \frac{1}{n} \log \left(P(b_n|b_{n-1}b_{n-2} \cdots b_1) \right)$$

and

$$-\frac{1}{n} \log \left(P(b_1|b_0 b_{-1} b_{-2} \cdots) \right) - \frac{1}{n} \log \left(P(b_2|b_1 b_0 b_{-1} \cdots) \right) - \frac{1}{n} \log \left(P(b_3|b_2 b_1 b_0 \cdots) \right)$$

converge to the same thing. You do this by choosing your $X_{i,j}$s in such a way that one sequence is the diagonal and the other is the limit of the columns. To get you started, here's the way the first 9 entries in the matrix should look:

$$\begin{pmatrix} -\log\left(P(b_1)\right) & -\log\left(P(b_2)\right) & -\log\left(P(b_3)\right) & \cdots \\ -\log\left(P(b_1|b_0)\right) & -\log\left(P(b_2|b_1)\right) & -\log\left(P(b_3|b_2)\right) & \cdots \\ -\log\left(P(b_1|b_{-1}b_0)\right) & -\log\left(P(b_2|b_0 b_1)\right) & -\log\left(P(b_3|b_1 b_2)\right) & \cdots \\ \vdots & \vdots & \vdots & \ddots \end{pmatrix}.$$

(Notice that the sequence $(X_{n,n})$ on the diagonal is exactly the thing you are trying to establish Cesàro convergence of.) To establish essential boundedness, one uses Theorem 355, with $a_i = |\Lambda| 2^{-i}$. The idea is that in order for $-\log\left(P(b_n|b_{n-1}b_{n-2}\cdots b_1)\right)$ to be between i and $i+1$ (indeed, in order for it to be at least i), b_n has to be a letter which, given b_{n-1}, b_{n-2}, etc. has probability at most 2^{-i}. There can't be more than $|\Lambda|$ such letters. But note there's some subtlety here. In showing that $(X_{n,n})$ is essentially bounded, for example, it isn't sufficient for the hypotheses of Theorem 355 that $P\left(i \leq -\log P(b_n|b_1 b_2 \cdots b_{n-1}) < i+1\right) < a_i$; you have to remember to condition on $(X_{i,i})_{i=1}^{n-1}$, so what you are actually showing is that for a.e. $b = (b_i)_{i=-\infty}^{\infty}$,

[74] We assume, as usual, the domain space to be $\Omega = \Lambda^{\mathbf{Z}}$.

$$P\Big(i \leq - \log P(b_n|b_1 b_2 \cdots b_{n-1}) < i + 1 \Big| - \log P(b_{n-1}|b_1 b_2 \cdots b_{n-2}),$$

$$- \log P(b_{n-2}|b_1 b_2 \cdots b_{n-3}), \ldots, - \log P(b_2|b_1), - \log(b_1)\Big)(b) < a_i. \quad (1)$$

(Remember, we use $P(b_n|b_1 b_2 \cdots b_{n-1})$ for $P(Y_n = b_n|Y_i = b_i, 1 \leq i \leq n - 1)$.)

Sketch of proof.

374. Exercise. Assuming the limit to exist, show that $H(x) = H(Tx)$ a.e., where T is the shift. Conclude that in the ergodic case, H is constant a.e. •

Let $X_{i,j}(x) = - \log \big(P(Y_j = x_j|Y_t = x_t, j - i + 1 \leq t \leq j - 1)\big)$. Also let $X_j(x) = \lim_{i \to \infty} X_{i,j}(x) = - \log \big(P(Y_j = x_j|Y_t = x_t, t \leq j - 1)\big)$. (This limit exists a.e. by the bounded martingale convergence theorem.) We must show:

(a) the columns form a stationary process;
(b) $(X_{i,i})$ and (X_i) are essentially bounded a.e.

In order to show (a), it suffices to show that for any $n, m \in \mathbf{N}$, any measurable sets $A_{i,j}, 1 \leq i \leq n, 1 \leq j \leq m$, and any $z \in \mathbf{N}$,

$$P(X_{i,j} \in A_{i,j}, 1 \leq i \leq n, 1 \leq j \leq m) = P(X_{i,z+j} \in A_{i,j}, 1 \leq i \leq n, 1 \leq j \leq m).$$

That is to say, we must show that

$$P\Big(- \log \big(P(Y_j = x_j|Y_t = x_t, j - i + 1 \leq t \leq j - 1)\big) \in A_{i,j},$$

$$1 \leq i \leq n, 1 \leq j \leq m\Big)$$

$$= P\Big(- \log \big(P(Y_{z+j} = x_{z+j}|Y_t = x_t, z + j - i + 1 \leq t \leq z + j - 1)\big) \in A_{i,j},$$

$$1 \leq i \leq n, 1 \leq j \leq m\Big).$$

375. Exercise. Denote the events of the foregoing equality by B_1 and B_2, respectively. Establish the equality by showing that $B_2 = T^{-z}B_1$, where T is the shift.

As to (b), we consider $(X_{i,i})$ first. The idea is to apply Theorem 355, supplemented with the following exercise.

376. Exercise. Establish (1), and conclude that $(X_{i,i})$ is essentially bounded a.e. *Hint: first show that* $P\Big(i \leq - \log P(Y_n = b_n|Y_i = b_i, 1 \leq i \leq n - 1) < i + 1\Big|Y_1, \ldots, Y_{n-1}\Big)(b) < a_i$ *for a.e. b (see the "Idea" section for how), then apply Exercise 126 (b).*

Showing that (X_i) is essentially bounded a.e. is similar. Now by the conclusion of the square lemma,

$$\lim_{n\to\infty} -\frac{1}{n} \log P(Y_t = x_t, 1 \le t \le n)$$

$$= \lim_{n\to\infty} -\frac{1}{n} \log \left(\prod_{i=1}^{n} P(Y_i = x_i | Y_t = x_t, 1 \le t \le i-1) \right)$$

$$= \lim_{n\to\infty} \frac{1}{n} \sum_{i=1}^{n} -\log \left(P(Y_i = x_i | Y_t = x_t, 1 \le t \le i-1) \right)$$

$$= \lim_{n\to\infty} \frac{1}{n} \sum_{i=1}^{n} X_{i,i}(x)$$

exists a.e., as required.   □

377. Comment. Like the Birkhoff theorem, versions of the Shannon–McMillan–Breiman theorem can be proved in various other circumstances, e.g. non-ergodic or asymptotically mean stationary systems. We refer the reader to Gray (1991, Chapter 3) for formulations, proofs and a wealth of historical information.

4.3. Entropy of a stationary process

This is a very short but important section in which we define the entropy of a process (see Kolmogorov 1958) and show that it is an isomorphism invariant.

378. Definition. The *entropy* $H\big((Y_i)\big)$ of an ergodic stationary process $(Y_i)_{i=-\infty}^{\infty}$ is the almost everywhere constant value of the function $H(x)$ in the Shannon–McMillan–Breiman theorem. In the general, possibly non-ergodic case, the entropy of $(Y_i)_{i=-\infty}^{\infty}$ is $E(H)$.

379. Corollary. *The entropy of* $(Y_i)_{i=-\infty}^{\infty}$ *is* $E\big(-\log P(Y_0|past)\big)$.

Proof. In the ergodic case, $H(x)$ is constant and equal to the limit of the Cesàro averages of the stationary process (X_i). Therefore, $H(x) = E(X_i)$ for any i. Taking $i = 0$, $H(x) = E(X_0)$, where $X_0(x) = -\log P(Y_0 = x_0 | Y_t = x_t, t < 0) = -\log P(Y_0|past)(x)$. (See the proof of Shannon–McMillan–Breiman.)

380. Exercise. Do the non-ergodic case.   □

381. Corollary. *A process has zero entropy if and only if the past determines the present (with probability 1).*

382. Comment. For an ergodic process, notice that the Shannon–McMillan–Breiman theorem implies (indeed, using just convergence in measure) that for all $\epsilon > 0$ it is the case that for all sufficiently large n, for every x outside an exceptional set of measure at most ϵ, the word $x_1 x_2 \cdots x_n$ has exponential size within ϵ of the entropy of the process.

383. Definition. In the ergodic case, by an ϵ-*reasonable name* we mean a word whose exponential size lies within ϵ of the entropy of the process.

384. Comment. In the ergodic case, the Shannon–McMillan–Breiman theorem implies that for large n, after removing a small set, all words of length n have approximately the same exponential size. Therefore (again after removal of a small set) the exponential number of words is approximately the entropy of the process.

385. Theorem. *The function that takes a process to its entropy is $\bar{d}$-continuous.*

386. Exercise. Prove the theorem. *Hint: the number of words of length n that are dbar-close to a given word is exponentially small; use the fact that the two processes admit of a coupling that usually spits out a pair of words close in dbar.*   □

387. Comment. It is important to note that the ratio of the probabilities of two ϵ-reasonable names can be arbitrarily large; it's the exponential sizes that are close.

388. Theorem. *Suppose that $(X_i)_{i=-\infty}^{\infty}$ is a stationary process and $(Y_i)_{i=-\infty}^{\infty}$ is a factor. Then the entropy of $(X_i)_{i=-\infty}^{\infty}$ is at least as great as the entropy of $(Y_i)_{i=-\infty}^{\infty}$.*

Idea of proof. In the ergodic case, adapt the argument of Theorem 365, utilizing Comment 382.

389. Exercise. Complete the proof.   □

We omit the proof of the following immediate corollary.

390. Corollary. *Any two isomorphic processes have the same entropy.*

4.4. An abstraction: partitions of 1

In the last section we introduced the entropy of a process, but we didn't do it in the normal way. The more usual way is to first define the entropy of a finite partition and then do a bunch of other stuff we'll get into later. All that matters in computing the entropy of a finite partition are the measures of the cells. In this section we consider finite partitions stripped of all but that information; we call these *partitions of 1*.

391. Definition. **p** is a *finite ordered partition of 1* if there is some n for which $\mathbf{p} = (p_1, p_2, \ldots, p_n)$, where $p_i \geq 0$, $1 \leq i \leq n$ and $\sum_{i=1}^n p_i = 1$. We refer to the n coordinates as the *cells* of the partition.

392. Comment. When there is no danger of confusion, we may simply say that **p** is a *partition of 1*.

393. Definition. Let $\mathbf{p} = (p_1, p_2, \ldots, p_n)$ be a finite ordered partition of 1. The *partition entropy* (denoted $H(\mathbf{p})$) of **p** is the entropy of any independent process $(X_i)_{i=-\infty}^{\infty}$ on an n-letter alphabet $\Lambda = \{\lambda_1, \ldots, \lambda_n\}$ with $P(X_0 = \lambda_i) = p_i$, $1 \leq i \leq n$.

394. Exercise. Show that the partition entropy of a partition of 1 is well defined.

395. Theorem. *The partition entropy of* $\mathbf{p} = (p_1, \ldots, p_n)$ *is* $\sum_{i=1}^n -p_i \log(p_i)$.

Idea of proof. Either apply Corollary 379 directly, or for a more intuitive proof, consider that the typical word of length n has approximately np_i occurrences of λ_i, $1 \leq i \leq n$, whence its probability is something like $\prod_{i=1}^n p_i^{np_i} = 2^{n(p_1 \log(p_1) + \cdots + p_n \log(p_n))}$. □

396. Comment. If (X, μ) is a probability space, where $X = \{x_1, \ldots, x_k\}$ is finite, then we may speak of the partition entropy of the measure μ. By this we mean the partition entropy of $\mathbf{p} = \big(\mu(x_1), \mu(x_2), \ldots, \mu(x_k)\big)$.

397. Definition. Let $\mathbf{p} = (p_1, \ldots, p_n)$ and $\mathbf{q} = (q_1, \ldots, q_m)$ be partitions of 1. By *interpolate* **p** *into the first cell of* **q** we mean "construct the $(n + m - 1)$-cell partition of 1 $\mathbf{r} = (q_1 p_1, q_1 p_2, \ldots, q_1 p_n, q_2, \ldots, q_m)$". By *interpolate* **p** *into the second cell of* **q** we mean "construct the $(n + m - 1)$-cell partition $\mathbf{r} = (q_1, q_2 p_1, q_2 p_2, \ldots, q_2 p_n, q_3, \ldots, q_m)$". Similarly for cells 3 through m.

398. Exercise. (*The conditioning property.*) Let **p** and **q** be partitions of 1 and let q_1 be the size of the first cell of **q**. If **p** is interpolated into the first cell of **q** then the resulting partition **r** has entropy $H(\mathbf{r}) = H(\mathbf{q}) + q_1 H(\mathbf{p})$.

399. Definition. Let $\mathbf{p} = (p_1, \ldots, p_n)$ and $\mathbf{q} = (q_1, \ldots, q_m)$ be partitions of 1. By a *join*, or *joining*, of **p** and **q** we mean an $n \times m$ matrix $\mathbf{a} = (a_{ij})$ of non-negative reals satisfying the following:

(a) $\sum_{j=1}^m a_{ij} = p_i$, $1 \leq i \leq n$;
(b) $\sum_{i=1}^n a_{ij} = q_j$, $1 \leq j \leq m$.

The *independent* join is given by $a_{ij} = p_i q_j$.

400. Comment. Notice that a join of two partitions of 1 can be viewed as a coupling of the obvious corresponding measures on $\{1, \ldots, n\}$ and $\{1, \ldots, m\}$, respectively.

401. Important comment. In practice, we want to consider a join of two partitions of 1 to itself be a partition of 1. Therefore we say $\mathbf{r} = (r_1, \ldots, r_t)$ is a join of $\mathbf{p}$ and $\mathbf{q}$, provided there is some understood bijection π taking the index set $\{1, 2, \ldots, t\}$ of R to the index set $\{1, 2, \ldots, n\} \times \{1, 2, \ldots, m\}$ of the canonical join $\mathbf{a}$ such that $r_i = a_{\pi(i)}$, $1 \leq i \leq t$. This allows one to speak meaningfully about the join of two joins, for example, or about the entropy of a join. Although this may be an abuse of notation, for the most part we'll completely suppress reference to such technical details.

402. Exercise. Show that the entropy of the independent join of partitions of 1 $\mathbf{p}$ and $\mathbf{q}$ is $H(\mathbf{p}) + H(\mathbf{q})$. *Hint: interpolate $\mathbf{p}$ into each of the original cells of $\mathbf{q}$ in turn. Use the conditioning property.*

403. Exercise. Let $\mathbf{p}^{(i)} = (p_{i,1}, p_{i,2}, \ldots, p_{i,r})$ be ordered partitions of 1 having the same number of cells r, $1 \leq i \leq k$, and let $a_1 + a_2 + \cdots + a_k = 1$, where $a_i \geq 0$, $1 \leq i \leq k$. Let $p_j = a_1 p_{1,j} + a_2 p_{2,j} + \cdots + a_k p_{k,j}$, $1 \leq j \leq r$. Then $\mathbf{p} = a_1 \mathbf{p}^{(1)} + a_2 \mathbf{p}^{(2)} + \cdots + a_k \mathbf{p}^{(k)} = (p_1, p_2, \ldots, p_r)$ is an ordered partition of 1.

404. Definition. A partition of 1 $\mathbf{p}$ constructed from partitions $\mathbf{p}^{(i)}$ as in the previous exercise is called a *convex combination* of the $\mathbf{p}^{(i)}$.

405. Comment. More generally, if $v_1 \ldots, v_k$ are elements of any vector space and $a_1 + a_2 + \cdots + a_k = 1$, where $a_i \geq 0$, $1 \leq i \leq k$, then $v = \sum_{i=1}^{k} a_i v_i$ is a convex combination of the v_i. This construction can apply to random variables or probability measures.

406. Theorem. *Suppose $\mathbf{p} = a_1 \mathbf{p}^{(1)} + a_2 \mathbf{p}^{(2)} + \cdots + a_k \mathbf{p}^{(k)}$ is a convex combination of partitions of 1. Then $H(\mathbf{p}) \geq \sum_{i=1}^{k} a_i H(\mathbf{p}^{(i)})$.*
Sketch of proof. Write $\mathbf{p}^{(i)} = (p_{i,1}, p_{i,2}, \ldots, p_{i,r})$ and $\mathbf{p} = (p_1, p_2, \ldots, p_r)$, where $p_j = a_1 p_{1,j} + a_2 p_{2,j} + \cdots + a_k p_{k,j}$, $1 \leq j \leq r$.

407. Exercise. Show that the function $f(x) = -x \log_2 x$ is concave on $(0, 1)$. *Hint: it suffices to show that $f'' < 0$ on that interval.*
Now by Jensen's inequality, one has

$$H(\mathbf{p}) = \sum_{j=1}^{r} f(p_j) = \sum_{j=1}^{r} f\left(\sum_{i=1}^{k} a_i p_{i,j}\right) \geq \sum_{j=1}^{r} \sum_{i=1}^{k} a_i f(p_{i,j})$$

$$= \sum_{i=1}^{k} a_i \sum_{j=1}^{r} f(p_{i,j}) = \sum_{i=1}^{k} a_i H(\mathbf{p}^{(i)}). \qquad \square$$

408. Theorem. *Let* **p** *and* **q** *be ordered partitions of 1. No joining of* **p** *and* **q** *has entropy greater than that of the independent joining.*

Sketch of proof. Let $\mathbf{p} = (p_1, \ldots, p_n)$, $\mathbf{q} = (q_1, \ldots, q_m)$ and let $\mathbf{a} = (a_{i,j})$ be the canonical join. For $1 \leq i \leq n$ put $\mathbf{a}_i = (\frac{a_{i,1}}{p_1}, \frac{a_{i,2}}{p_1}, \ldots, \frac{a_{i,m}}{p_1})$.

409. Exercise. Each $\mathbf{a}_i$ is a partition of 1 and $\mathbf{q} = \sum_{i=1}^{n} p_i \mathbf{a}_i$.   •

410. Exercise. One can get back to a partition of 1 equivalent to $\mathbf{a}$ by interpolating, in turn, each $\mathbf{a}_i$ into the ith cell of $\mathbf{p}$. Conclude that $H(\mathbf{a}) = H(\mathbf{p}) + \sum_{i=1}^{n} p_i H(\mathbf{a}_i)$.   •

The proof is completed by observing that $H(\mathbf{p}) + \sum_{i=1}^{n} p_i H(\mathbf{a}_i) \leq H(\mathbf{p}) + H(\mathbf{q})$ (which is the entropy of the independent join) by the previous exercise and Theorem 406.   □

411. Definition. Let $\mathbf{p}$ and $\mathbf{q}$ be ordered partitions of 1 and let $\mathbf{a}$ be a joining of $\mathbf{p}$ and $\mathbf{q}$. Then the *entropy of* $\mathbf{p}$ *given* $\mathbf{q}$ *relative to* $\mathbf{a}$ is the quantity $H(\mathbf{p}|_{\mathbf{a}}\mathbf{q}) = H(\mathbf{a}) - H(\mathbf{q})$.

412. Comment. Observe that $H(\mathbf{p}|_{\mathbf{a}}\mathbf{q}) \leq H(\mathbf{p})$ as an immediate consequence of Theorem 408.

4.5. Measurable partitions, entropy of measure-preserving systems

In this section we consider real partitions (instead of their abstractions, as in the last section), and show that our definition of entropy coincides with the classical one via partitions. Finally we define the entropy of a general measure-preserving system.

413. Exercise. Let $(\Omega, \mathcal{A}, \mu)$ be a probability space and let $A = \{A_1, A_2, \ldots, A_n\}$ be a measurable partition of Ω. Then $\mathbf{p}_A = (\mu(A_1), \mu(A_2), \ldots, \mu(A_n))$ is an ordered partition of 1.

414. Definition. Let $A = \{A_1, A_2, \ldots, A_n\}$ be a measurable partition; assume $\mu(A_1) \leq \mu(A_2) \leq \cdots \leq \mu(A_n)$. We denote by $\mathbf{p}_A$ the partition of 1 $(\mu(A_1), \mu(A_2), \ldots, \mu(A_n))$. The *entropy* of A is given by $H(A) = H(\mathbf{p}_A)$.

415. Comment. Let $A = \{A_1, A_2, \ldots, A_n\}$ be a measurable partition of a probability space $(\Omega, \mathcal{A}, \mu)$. If one defines $I(x)$ by the rule $x \in A_{I(x)}$ (so that I just picks out what cell one is in), then one can check that $H(A) = E(-\log(\mu(A_I)))$. That is to say "entropy of a partition is the expectation of $-\log$ of the probability of the cell you're in".

416. Definition. Let A and B be finite measurable partitions. The *entropy of* *A* *given* *B* is the quantity $H(A|B) = H(A \vee B) - H(B)$.

417. Comment. Let $A = \{A_1, A_2, \ldots, A_n\}$ and $B = \{B_1, B_2, \ldots, B_m\}$ be finite measurable partitions of a probability space $(\Omega, \mathcal{A}, \mu)$ and let $C = \{C_{i,j} : 1 \leq i \leq n, 1 \leq j \leq m\}$ be their superimposition, $C = A \vee B$. Here $C_{i,j} = A_i \cap B_j$. Then $\mathbf{p}_C$ is a join of $\mathbf{p}_A$ and $\mathbf{p}_B$.

418. Exercise. $H(A|B) = H(\mathbf{p}|_\mathbf{r}\mathbf{q})$, where $\mathbf{p} = \mathbf{p}_A$, $\mathbf{q} = \mathbf{p}_B$ and $\mathbf{r} = \mathbf{p}_{A \vee B}$.

419. Comment. One has $H(A|B) \leq H(A)$. (Cf. Comment 412.)

420. Definition. Let A and B be measurable partitions such that every cell of A is a union of cells of B. We say that A is a *subpartition* of B and that B is a *refinement* of A.

421. Theorem. *Let Q be a subpartition of R, and let P be another partition. Then $H(P|R) \leq H(P|Q)$.*

Idea of proof. Construct measurable partitions P', Q' and R' of $[0, 1]$ having the same numbers of cells as P, Q and R respectively, such that $\mu(P_i' \cap Q_j') = \mu(P_i \cap Q_j)$, $\mu(R_k') = \mu(R_k)$, and such that $Q_j' = R_{i_1}' \cup \cdots \cup R_{i_t}'$ if and only if $Q_j = R_{i_1} \cup \cdots \cup R_{i_t}$. In addition, arrange that each $P_i' \cap Q_j'$ is independent of the partition that R' induces on Q_j'. Show that $H(P|R) \leq H(P'|R') = H(P'|Q') = H(P|Q)$. (Use concavity of $-x \log_2 x$ in showing that $H(P \vee R) \leq H(P' \vee R')$.) $\quad\square$

422. Definition. Let X_i and Y_j be random variables taking values in a finite alphabet, $0 \leq i \leq n$, $0 \leq j \leq m$. By $H(X_i, 1 \leq i \leq n | Y_j, 0 \leq j \leq m)$ we mean $H(A|B)$, where A is the algebra generated by the X_i and B is the algebra generated by the Y_j.

423. Corollary. *Let $(X_i)_{i=-\infty}^{\infty}$ be a stationary process on a finite alphabet. Then*

(a) $H(X_n | X_i, 0 \leq i < n)$ is non-increasing in the variable n;
(b) $\frac{1}{n} H(X_i, 0 \leq i < n)$ is non-increasing in the variable n.

Sketch of proof. $H(X_n | X_i, 0 \leq i < n) = H(X_0 | X_i, -n \leq i < 0)$ by stationarity. Apply Theorem 421 to get (a). Now (b) follows from (a) easily. $\square$

The following theorem shows our definition of entropy to be equivalent to a more classical one, namely the right-hand side of the display below.

424. Theorem. *Let $(X_i)_{i=-\infty}^{\infty}$ be a stationary process on a finite alphabet. Then*

$$H\left((X_i)_{i=-\infty}^{\infty}\right) = \lim_{n \to \infty} H(X_n | X_i, 0 \leq i < n) = \lim_{n \to \infty} \frac{1}{n} H(X_i, 0 \leq i < n).$$

Sketch of proof. By Corollary 379, one has

$$H\big((X_i)_{i=-\infty}^{\infty}\big)$$

$$= E\big(-\log P(X_0|past)\big)$$

$$= \lim_{n\to\infty} E\big(-\log P(X_0|X_i, N \le i < 0)\big)$$

$$= \lim_{n\to\infty} \sum_{b_0,b_{-1},\ldots,b_{-n}\in\Lambda} P(b_0 b_{-1}\cdots b_{-n})\Big(-\log \frac{P(b_0 b_{-1}\cdots b_{-n})}{P(b_{-1}\cdots b_{-n})}\Big)$$

$$= \lim_{n\to\infty} \sum_{b_0,b_{-1},\ldots,b_{-n}\in\Lambda} P(b_0 b_{-1}\cdots b_{-n})$$

$$\Big(-\log P(b_0 b_{-1}\cdots b_{-n}) + \log P(b_{-1}\cdots b_{-n})\Big)$$

$$= \lim_{n\to\infty} \Big(\sum_{b_0,b_{-1},\ldots,b_{-n}\in\Lambda} -P(b_0 b_{-1}\cdots b_{-n})\log P(b_0 b_{-1}\cdots b_{-n})$$

$$- \sum_{b_{-1},\ldots,b_{-n}\in\Lambda} -P(b_{-1}\cdots b_{-n})\log \big(P(b_{-1}\cdots b_{-n})\big)\Big)$$

$$= \lim_{n\to\infty} \big(H(X_i, -n \le i \le 0) - H(X_i, -n \le i \le -1)\big)$$

$$= \lim_{n\to\infty} H(X_0|X_i, -n \le i < 0) = \lim_{n\to\infty} H(X_n|X_i, 0 \le i < n).$$

This takes care of the first equality. For the second, observe that

$$H(X_i, 0 \le i < n) = H(X_0) + H(X_1|X_0) + H(X_2|X_0, X_1) + \cdots$$
$$+ H(X_n|X_0, \ldots, X_{n-1}),$$

and use the fact that $x_n \to x$ implies $C \lim x_n \to x$.   □

425. Theorem. *Let $(X_i)_{i=-\infty}^{\infty}$ be a stationary process on a finite alphabet and let $m \in \mathbf{N}$. Then $H = H\big((X_i)_{i=-\infty}^{\infty}\big) = \frac{1}{m}\lim_{n\to\infty} H(X_i, n+1 \le i \le n+m|X_j, 0 \le j \le n)$.*

Sketch of proof.

$$\lim_{n\to\infty} H(X_i, n+1 \le i \le n+m|X_j, 0 \le j \le n)$$
$$= \lim_{n\to\infty} H(X_i, 1 \le i \le m|X_j, -n \le j \le 0),$$

which exists by Theorem 421. Call the limit L. Our job is to show that $L = mH$. Now we mimic the preceding proof. For $k \in \mathbf{N}$,

$$\frac{1}{k}H(X_i, 1 \le i \le km) = \frac{1}{k}\Big(H(X_i, 1 \le i \le m)$$
$$+ H(X_i, m+1 \le i \le 2m | X_i, 1 \le i \le m)$$
$$+ H(X_i, 2m+1 \le i \le 3m | X_i, 1 \le i \le 2m) + \cdots$$
$$\cdots + H(X_i, (k-1)m+1 \le i \le km | X_i, 1$$
$$\le i \le (k-1)m)\Big).$$

426. Exercise. As $k \to \infty$, the left-hand side goes to mH, the right-hand side to L. □

427. Definition. Let $(\Omega, \mathcal{A}, \mu, T)$ be a measure-preserving system, and let P be a finite measurable partition of Ω. We denote by $H(P, T)$ the entropy of the (P, T) process.

The *entropy* of $(\Omega, \mathcal{A}, \mu, T)$ is the supremum, over all finite measurable partitions P, of $H(P, T)$. We denote the entropy of $(\Omega, \mathcal{A}, \mu, T)$ by $H(\Omega, \mathcal{A}, \mu, T)$, or simply by $H(T)$.

428. Comment. If P and Q are two finite generators then $H(P, T) = H(Q, T)$. The reason for this is that the (P, T) process is isomorphic to the (Q, T) process. In fact, if a transformation has a finite generator P then $H(T) = H(P, T)$, since for any other finite partition Q, $P \vee Q$ is a generator too, so $H(Q, T) \le H(P \vee Q, T) = H(P, T)$.

4.6. Krieger finite generator theorem

We won't have to use the material in this section, but we don't think the reader should skip it because the methods used herein are very important. We start out with a couple of technical exercises and then move to the Krieger finite generator theorem, which is a deep result with a demanding proof. The reader who is patient enough to work through the details will be rewarded in the remainder of the book.

429. Exercise. Given an upper bound on the number of cells, show that the entropy of the factor generated by a partition is continuous in the variation metric. That is to say: let $Q \in \mathbf{N}$ and $\epsilon > 0$. Show there exists $\delta = \delta(\epsilon, Q) > 0$ such that if $(\Omega, \mathcal{A}, \mu, T)$ is any system and $P = \{P_1, \ldots, P_Q\}$, $F = \{F_1, \ldots, F_Q\}$ are any Q-cell measurable partitions of Ω satisfying $v(H, F) < \delta$ (see Definition 342) then $|H(P, T) - H(F, T)| < \epsilon$. *Hint: choose δ so that both $-\delta \log \delta$ and $\delta \log Q$ are small relative to ϵ.*

430. Exercise. Let $k \in \mathbf{N}$ and let $k' < k$ be a real number. Take $\Lambda = \{1, 2, \ldots, k\}$. There exists $u \in \mathbf{N}$ such that if r is any word on Λ of length u then for every $J \in \mathbf{N}$ it is the case that for all sufficiently large n,

$\left|\left\{w : w \text{ is a word on } \Lambda \text{ of length } n - J \text{ not having } r \text{ as a subword}\right\}\right| > (k')^n.$

Hint: take u large enough that $k^u - 1 > (k')^u$.

431. Comment. We're now ready to state the Krieger finite generator theorem. For the proof, we're going to provide more detail than the reader will be used to seeing in this book. The reason for this is that this proof is a paradigm for many of the constructions to come in later chapters. We view it as *absolutely essential* that the details be worked through rigorously.

432. Theorem. *(Krieger finite generator theorem; Krieger 1970.) Let $k \in \mathbf{N}$ and suppose an ergodic system $(\Omega, \mathcal{A}, \mu, T)$ has entropy $h < \log k$. Then $(\Omega, \mathcal{A}, \mu, T)$ has a k-cell generator. (See Definition 154 for the definition of "generator".)*

Idea of proof. Let (Q_n) be a nested sequence of finite partitions that separates points. We will look at a Rohlin tower, then a much bigger one, then a much bigger one, etc. Start with a large m. Let $n \gg m$. We will choose the first tower of height n, so that the base is independent of the partition according to n names of Q_m. By Shannon–McMillan–Breiman, there are enough n-names on a k-letter alphabet to label the rungs of all reasonable columns with the letters $B_1, B_2, \ldots, B_k$ in such a way that the reasonable columns are distinguished from each other, because the number of reasonable columns is smaller than the number of such names. Carry out such a labeling. If you know the $\{B_1, B_2, \ldots, B_k\}$-name of a column containing the base of the tower, you probably know the Q_m-name of that column. The only reason you might turn out to be wrong is that you might be looking at an unreasonable column.

This is just the first stage. We will end up altering our labeling $B_1, B_2, \ldots, B_k$ over and over again, to accommodate approximation of Q_m for larger and larger m, however altering the labeling less and less so that it stabilizes, and so that starting with a random point x which is labeled at some stage, the label of x is unlikely to change in subsequent alterations. In the end, we will be able to tell all of the Q_i names from our $\{B_1, \ldots, B_k\}$-name, and hence we will be able to tell what point we are, from our $\{B_1, \ldots, B_k\}$-name. Hence we will know that $\{B_1, \ldots, B_k\}$ generates.

There is a complication. We will have to know whenever we enter one of these towers from just looking at the $\{B_1, \ldots, B_k\}$-name. If we saw an infinite $\{B_1, \ldots, B_k\}$-name but had no idea which coordinates were in the base of the tower, we would be quite confused. To handle this, we need to reserve certain words to label the bases of the towers, so that we know when we are in one. You can arrange that so few words are reserved that no damage is done to the proof. Use labels like 21111112 for the first tower and 211111112 for the second, etc. (the reader needs to figure out how long these labels need to be so

that the restriction that these names cannot be used except for labeling the base
will not cause damage).

More details

1. Start off with Q_{10} (10 chosen so that Q_{10} has nearly full entropy). Pick n_1
   large enough so that Shannon–McMillan–Breiman kicks in for Q_{10}-names
   by time n_1. Pick a Rohlin tower of height n_1 whose base is independent of
   the partition of Q_{10} names of length n_1. Label the reasonable columns with
   distinct $\{B_1, \ldots, B_k\}$-names.
2. Pick your second tower to be much bigger than your first, and as before pick
   your base independent of Q_{1000}-names of length the height of the tower
   (1000 chosen so that Q_{1000} has very nearly full entropy). The $\{B_1, \ldots, B_k\}$
   process defined at the previous stage already captures a great deal of
   entropy, because $\{B_1, \ldots, B_k\}$-names almost determine which atom of Q_{10}
   you are in. Prove that there are enough $\{B_1, \ldots, B_k\}$-names so that, since
   our entropy is strictly less than $\log k$, it is possible to let the top $\frac{3}{4}$ of
   the rungs remain as they are and change the bottom $\frac{1}{4}$ of the rungs to
   $\{B_1, \ldots, B_k\}$-names in such a way as to distinguish all reasonable Q_{1000}
   columns.
3. Continue, except replace the numbers $\frac{3}{4}$ and $\frac{1}{4}$ with $\frac{7}{8}$ and $\frac{1}{8}$ at the next
   stage, then with $\frac{15}{16}$ and $\frac{1}{16}$ in the following stage, etc.
4. Here is why the final $\{B_1, \ldots, B_k\}$-name separates points. At the end of
   the first stage you probably know what atom of Q_{10} you are in and there
   is not much more than a $\frac{1}{4}$ chance that the information will be damaged.
   Continue this reasoning at higher stages using Borel–Cantelli to show that
   information is damaged only finitely many times (with probability 1).

Sketch of proof.

Put $\Lambda = \{1, 2, \ldots, k\}$. Choose a real number $k' < k$ such that $h < \log k'$.
Now choose u as in Exercise 430 and let r be the word of length u consisting
entirely of 1s. Also, for any $i \in \mathbf{N}$, let r_i be the word of length $u + i + 1$
consisting of a 2 followed by $u + i - 1$ consecutive 1s then another 2. Notice
that every r_i contains r as a subword. For $n \in \mathbf{N}$, let $W_n^{(i)}$ consist of all words
on Λ of length n that begin with r_i but contain no other occurrence of r. By
the conclusion of Exercise 430 (with $J = u + i + 1$), $|W_n^{(i)}| > (k')^n$ for all
sufficiently large n.

Let $(P^{(i)})_{i=1}^{\infty}$ be an increasing sequence of finite partitions of Ω that separates points. Let $h' = \lim_{i \to \infty} H(P^{(i)}, T) \le h.$[75] Choose i_1 such that

---

[75] One actually has $h' = h$, however our proof does not make use of this fact.

$h' - H(P^{(i_1)}, T) < \epsilon_1 = \frac{1}{10}(\log k' - h')$. Letting now Q_1 be the number of cells in $P^{(i_1)}$, let $\delta_1 = \min\{\frac{1}{3}\delta(\epsilon_1, Q_1), \frac{1}{6}\}$ (cf. Exercise 429).

Choose $n_1 > \frac{|r_1|}{\delta_1}$ such that $|W_{n_1}^{(1)}| > (k')^{n_1}$ and such that

$$P(x_0 x_1 x_2 \ldots x_{n_1-1} \text{ is an } \epsilon_1\text{-reasonable } P^{(i_1)}\text{-name}) > 1 - \delta_1.$$

Let $\mathcal{A}_1$ be the algebra generated by the $P^{(i_1)}$-names of length n_1, and let S_1 be the base of a Rohlin tower of height n_1 with an error set E_1 having measure at most δ_1 such that S_1 is independent of $\mathcal{A}_1$.

433. Exercise. Show that there are at most $2^{n_1(H(P^{(i_1)},T)+\epsilon_1)} < (k')^{n_1}$ reasonable $P^{(i_1)}$-names of length n_1. Conclude that, neglecting a set $D_1 \subset S_1$ with $\mu(D_1) < \frac{\delta_1}{n_1}$, one can partition S_1 into fewer than $(k')^{n_1}$ cells $C_1, \ldots$ such that any two members of the same cell have the same $P^{(i_1)}$-name on $\{0, 1, \ldots, n_1 - 1\}$. •

Let π be any injective mapping from the cells $C_1, \ldots$ into $W_{n_1}^{(1)}$. (There are such mappings since $|W_n^{(1)}| > (k')^n$.) We now use π to create a k-cell ordered partition of Ω as follows: if $x \in C_i$ and $\pi(C_i) = a_0 a_1 \ldots a_{n_1-1}$, put $x \in B_{a_0}^{(1)}$, $Tx \in B_{a_1}^{(1)}$, $T^2 x \in B_{a_2}^{(1)}, \ldots, T^{n_1-1}x \in B_{a_{n_1-1}}^{(1)}$. Finally, put $E_1 \cup \left(\bigcup_{i=0}^{n_1-1} T^i D_1 \right)$ into $B_2^{(1)}$. Denote the resulting partition by $B^{(1)}$.

Let $M_1 = E_1 \cup \left(\bigcup_{i=0}^{n_1-1} T^i D_1 \right) \cup \bigcup_{i=0}^{|r_1|-1} T^i S_1$. Note that $\mu(M_1) < 3\delta_1 \leq \frac{1}{2}$.

434. Exercise. Show that if $x \notin M_1$, then by knowing $x_{-n_1} x_{-n_1+1} \cdots x_{n_1}$ (as a $B^{(1)}$-name), one can figure out which cell of $P^{(i_1)}$ x is in. To be more precise: write $P^{(i_1)}$ as an ordered partition $\{G_1, \ldots, G_{Q_1}\}$. There is an ordered partition $F^{(1)} = \{F_1^{(1)}, \ldots, F_{Q_1}^{(1)}\}$ that is measurable with respect to the algebra generated by $\{T^i B_j^{(1)} : -n_1 \leq i \leq n_1, 1 \leq j \leq k\}$ and such that if $x \in G_m$ and $x \notin M_1$ then $x \in F_m^{(1)}$, $1 \leq m \leq Q_1$. *Hint: locate x in the tower by finding an occurrence of r_1 in the word $x_{-n_1+1} x_{-n_1+2} \cdots x_{-1} x_0$. If this works, you know what cell of $P^{(i_1)}$ x is in, so you put x in the corresponding cell of $F^{(1)}$. If it fails, you know $x \in M_1$. Put such x in $F_2^{(1)}$.* •

Note by choice of δ_1, $H(B^{(1)}, T) \geq H(F^{(1)}, T) > H(P^{(i_1)}, T) - \epsilon_1 > h' - 2\epsilon_1$.

Now choose $i_2 > i_1$ such that $h' - H(P^{(i_2)}, T) < \epsilon_2 = \frac{1}{100}(\log k' - h')$. Let $P_B^{(i_2)}$ be the superimposition of $B^{(1)}$ and $P^{(i_2)}$. Letting now Q_2 be the number of cells in $P_B^{(i_2)}$, let $\delta_2 = \min\{\frac{1}{4}\delta(\epsilon_2, Q_2), \frac{1}{16}\}$ (cf. Exercise 429).

Now, for all sufficiently large n divisible by 4, the following hold:

(1) $|W_{\frac{n}{4}}^{(2)}| > (k')^{\frac{n}{4}}$;

(2) $n > \frac{|r_2|}{\delta_2}$;

(3) $P(x_0 x_1 \cdots x_{\frac{3n}{4}-1}$ is an ϵ_2-reasonable $B^{(1)}$-name) $> 1 - \delta_2$;

(4) $P(x_0 x_1 \cdots x_{\frac{3n}{4}-1}$ is an ϵ_2-reasonable $P_B^{(i_2)}$-name) $> 1 - \delta_2$;

(5) $n > 2n_1$.

Choose n_2 satisfying (1)–(5) and let $\mathcal{A}_2$ be the algebra generated by the $P_B^{(i_2)}$-names of length $\frac{3}{4}n_2$.[76] Let S_2 be the base of a Rohlin tower of height n_2 with an error set E_2 having measure at most δ_2 such that every rung of the tower is independent of $\mathcal{A}_2$. (This is possible by Exercise 189. The reader should make note, however, that we don't actually need every rung to be independent; in fact only one particular rung needs to be so.) Notice that for every $B^{(1)}$-name w on $\{0, 1, \ldots, \frac{3}{4}n_2 - 1\}$, $\varphi(w)$ is a union of sets $\varphi(v)$, where each v is a $P_B^{(i_2)}$-name. If v_1 and v_2 are two ϵ_2-reasonable $P_B^{(i_2)}$-names on $\{0, 1, \ldots, \frac{3}{4}n_2 - 1\}$, write $v_1 \sim v_2$ if $\varphi(v_1) \subset \varphi(w)$ and $\varphi(v_2) \subset \varphi(w)$ for some ϵ_2-reasonable $B^{(1)}$-name w on $\{0, 1, \ldots, \frac{3}{4}n_2 - 1\}$.

435. Exercise. Show that the measure of $\varphi(w)$, where w is an ϵ_2-reasonable $B^{(1)}$-name on $\{0, 1, \ldots, \frac{3}{4}n_2 - 1\}$, is at most $2^{-\frac{3}{4}n_2(H(B^{(1)},T)-\epsilon_2)}$, and the measure of $\varphi(v)$, where v is a reasonable $P_B^{(i_2)}$-name of length $\frac{3}{4}n_2$, is at least $2^{-\frac{3}{4}n_2(H(P_B^{(i_2)},T)+\epsilon_2)}$. $\quad\bullet$

We now focus attention on the particular rung $T^{\frac{n_2}{4}}S_2$ of our Rohlin tower. In the next exercise, you are asked to show that, upon deleting the points of this rung giving rise to unreasonable names, on what's left there is an exponentially small number of represented $P_B^{(i_2)}$-names v in each equivalence class under $\sim$.

436. Exercise. There is a set $T^{\frac{n_2}{4}}D_2 \subset T^{\frac{n_2}{4}}S_2$ with $\mu(D_2) < \frac{2\delta_2}{n_2}$ having the property that for every $x \in T^{\frac{n_2}{4}}(S_2 \setminus D_2)$, letting $w = x_0 x_1 \cdots x_{\frac{3}{4}n_2}$ (as a $B^{(1)}$-name), there are no more than $2^{\frac{3}{4}n_2(2\epsilon_1 + 2\epsilon_2)}$ elements in the set $C(w)$ of $P_B^{(i_2)}$-names v on $\{0, 1, \ldots, \frac{3}{4}n_2 - 1\}$ for which there is some $y \in T^{\frac{n_2}{4}}(S_2 \setminus D_2)$ such that $v = y_0 y_1 \cdots y_{\frac{3}{4}n_2 - 1}$ (as a $P_B^{(i_2)}$-name) and $w = y_0 y_1 \cdots y_{\frac{3}{4}n_2 - 1}$ (as a $B^{(1)}$-name). $\quad\bullet$

Let x and w be as in the previous exercise. Since $|C(w)| < (k')^{\frac{n_2}{4}} < |W_{\frac{n_2}{4}}^{(2)}|$, there is an injective function $\pi_w : W_{\frac{n_2}{4}}^{(2)} \to C(w)$. We now use the maps π_w to modify the first partition $B^{(1)}$ as follows. If $y \in T^l(S_2 \setminus D_2)$, where $0 \le l < \frac{n_2}{4}$, let $x = T^{\frac{n_2}{4}-l}y \in T^{\frac{n_2}{4}}(S_2 \setminus D_2)$, write $w = x_0 x_1 \cdots x_{\frac{3}{4}n_2 - 1}$ (as a $B^{(1)}$-name), write $v = x_0 x_1 \cdots x_{\frac{3}{4}n_2 - 1}$ (as a $P_B^{(i_2)}$-name), write $a_0 a_1 \cdots a_{\frac{n_2}{4}-1} = \pi_w(v)$

---

[76] We mean the algebra whose atoms are the equivalence classes of the relation $x \sim y$ if and only if $x_0 x_1 \cdots x_{\frac{3}{4}n_2 - 1} = y_0 y_1 \cdots y_{\frac{3}{4}n_2 - 1}$ (as $P_B^{(i_2)}$-names).

and put $y \in B_{a_l}^{(2)}$. For any other point y, put $y \in B_i^{(2)}$ if and only if $y \in B_i^{(1)}$, $1 \le i \le k$.

437. Exercise. Let $V_2 \subset \bigcup_{i=1}^{n_2}(S_2 \setminus D_2)$ be the set of points that were modified in moving from $B^{(1)}$ to $B^{(2)}$. That is, $V_2 = \bigcup_{t \ne s} B_t^{(1)} \cap B_s^{(2)}$. Show that:

(a) $\mu(V_2) < \frac{1}{4}$, and

(b) $\mu\left(\bigcup_{i=-n_1}^{n_1} T^i V_2\right) < \frac{1}{2}$. *Hint: recall $n_2 > 2n_1$.*

Let $M_2 = E_2 \cup \left(\bigcup_{i=0}^{n_1-1} T^i D_2\right) \cup \bigcup_{i=0}^{|r_2|-1} T^i S_2$. Note that $\mu(M_2) < 4\delta_2 \le \frac{1}{4}$. •

438. Exercise. Show that if $x \notin M_2$, then by knowing $x_{-n_2}x_{-n_2+1} \cdots x_{n_2}$ (as a $B^{(2)}$-name), one can figure out which cell of $P^{(i_2)}$ x is in. (See Exercise 434 for clarification.) •

439. Exercise. Establish by induction that one may find, for all $j \in \mathbf{N}, i_j, n_j \in \mathbf{N}$, a measurable set M_j, and a k-cell ordered partition $B^{(j)}$ of Ω satisfying, for all $j > 1$:

(a) $h' - H(P^{(i_j)}, T) < \epsilon_j = \frac{1}{10^j}(\log k' - h')$;

(b) $\mu(M_j) \le 2^{-j}$;

(c) $H(B^{(j)}, T) > h' - 2\epsilon_j$;

(d) $i_j > i_{j-1}$;

(e) $v(B^{(j-1)}, B^{(j)}) \le 2^{-j}$;

(f) $\left(\bigcup_{i=-n_{j-1}}^{n_{j-1}} T^i V_j\right) < 2^{-j+1}$, where $V_j = \bigcup_{t \ne s} B_t^{(j-1)} \cap B_s^{(j)}$;

(g) if $x \notin M_j$, then by knowing $x_{-n_j}x_{-n_j+1} \cdots x_{n_j}$ (as a $B^{(j)}$-name), one can figure out which cell of $P^{(i_j)}$ x is in;

(h) $n_j > 2n_{j-1}$.

Hint: the above few paragraphs do exactly this for the case $j = 2$. •

440. Exercise. Show that $\lim_{j \to \infty} B_i^{(j)} = B_i$ exists a.e., $1 \le i \le k$.[77] *Hint: use (e) above and Borel–Cantelli.* •

We claim that the ordered partition $B = \{B_1, B_2, \ldots, B_k\}$ separates points. To see this, let R_1 be the set of points $x \in \Omega$ that lie in only finitely many of the sets $\bigcup_{i=-n_{j-1}}^{n_{j-1}} T^i V_j$, $j > 1$. By (f) above and Borel–Cantelli, $\mu(R_1) = 1$. Let R_2 be the set of points $x \in \Omega$ that lie in only finitely many of the sets M_j, $j \in \mathbf{N}$. Again by Borel–Cantelli, $\mu(R_2) = 1$. Finally let R_2 be a set of full measure such that for any $x, y \in R_2, x \ne y$, there is some $i \in \mathbf{N}$ such that $P^{(i)}$ separates x and y. Let $R = R_1 \cap R_2 \cap R_3$.

---

[77] That is to say, $\lim_{j \to \infty} 1_{B_i^{(j)}}(x)$ exists a.e., and the limit is clearly an indicator function. We let B_i be the set of points where the limit is 1.

Suppose that $x, y \in R$ with $x \neq y$. We claim that the B-names $(x_i)_{i=-\infty}^{\infty}$ and $(y_i)_{i=-\infty}^{\infty}$ are different. To see this, choose m so large that x and y are not in the "bad" sets $\bigcup_{i=-n_{j-1}}^{n_{j}-1} T^i V_j$ or M_j for any $j \geq m$, and such that $P^{(i_m)}$ separates x and y. By (g) above, $x_{-n_m} x_{-n_m+1} \cdots x_{n_m}$ and $y_{-n_m} y_{-n_m+1} \cdots y_{n_m}$ as $B^{(m)}$-names are different. But these are the same as $x_{-n_m} x_{-n_m+1} \cdots x_{n_m}$ and $y_{-n_m} y_{-n_m+1} \cdots y_{n_m}$ as B-names, so we're done. $\qquad\square$

441. Abbreviation of Proof. Make a big tower distinguishing reasonable columns in mean Hamming distance. Improve the approximation with a bigger tower by altering a few columns at the bottom. Label the bases of all towers with words reserved for this purpose.

442. Definition. Let $(X_i)_{i=-\infty}^{\infty}$ and $(Y_i)_{i=-\infty}^{\infty}$ be stationary processes on alphabets Λ_1 and Λ_2, respectively. Realize Y as a system $(Y, \mathcal{A}, \mu, T)$ and realize X as a system $(X, \mathcal{B}, \nu, S)$. Suppose there is a homomorphism $\pi :$ $X \to Y$ having the property that for every $\lambda \in \Lambda_2$, $\pi^{-1}\{y \in Y : y_0 = \lambda\}$ is a finite union of cylinder sets in X. Then $(Y_i)_{i=-\infty}^{\infty}$ is said to be a *coded factor* of $(X_i)_{i=-\infty}^{\infty}$, and the map taking a cylinder set in the preimage back to λ is a *code*.

443. The following lemma will be used in the next section.

Lemma. *Let $(X, \mathcal{A}, \mu, T)$ be an ergodic measure-preserving system and suppose that $Q = (A_1, A_2, \ldots)$ is a countable generating partition. Then $H(T) = \lim_{n \to \infty} H(Q_n, T)$, where $Q_n = (A_1, A_2, \ldots, A_{n-1}, \bigcup_{i=n}^{\infty} A_i)$.*

Sketch of proof. Obviously $H(T) \geq \lim_{n \to \infty} H(Q_n, T)$ (the limit exists, of course, as the sequence is monotone). Let P be an arbitrary finite measurable partition and let $\epsilon > 0$ be arbitrary. We will find an n such that $H(Q_n, T) \geq H(P, T) - \epsilon$, completing the proof.

Let δ be a very small number depending on ϵ and $|P|$ to be specified later. Since Q generates, the (infinite) Q-name of a point determines the P-name. Choose m so large that the Q-name of length $2m + 1$ centered at 0 of a point x determines the cell of P that x is in with probability at least $1 - \delta$. Next, choose n so large that $\mu(\bigcup_{i=n}^{\infty} A_i) < \frac{\delta}{2m+1}$. Then the probability that the Q_n-name of length $2m + 1$ centered at the origin determines which cell of P x is in is at least $1 - 2\delta$. Let B be the bad set where this code makes errors.

Next, choose a really big N such that Shannon–McMillan–Breiman has δ-kicked in for Q_n names of length N and P-names of length N and where Birkhoff ergodic has δ-kicked in for the bad set B.

444. Exercise. Show that the number of reasonable P-names of length N is no more than something like

$$2^{N(H(Q_n,T)+\delta)} \binom{N}{3\delta N} |P|^{3\delta N+2m}.$$

Then show that for small enough δ depending on ϵ and $|P|$, and big enough N, this is less than $2^{N(H(Q_n,T)+\epsilon)}$. Use this to finish the proof.   □

4.7. The induced transformation and fbar

This section can be skipped if desired. We show how a transformation on a measure space induces transformations on its subsets, and describe the relationship between the entropy of the original system and the entropy of the induced system.

445. Discussion. Let $(X, \mathcal{A}, \mu, T)$ be an ergodic measure-preserving system, and let A be a set of positive measure. Let $\mathcal{B}$ be the σ-algebra of subsets of B that is obtained by restricting $\mathcal{A}$ to subsets of A. Let ν be the probability measure on $\mathcal{B}$ defined by $\nu(B) = \frac{\mu(B)}{\mu(A)}$ for $B \in \mathcal{B}$. Define a transformation $S : A \to A$ by $Sx = T^i x$, where i is the least positive integer such that $T^i x \in A$.

446. Exercise. Show that $(A, \mathcal{B}, \nu, S)$ is a measure-preserving system on a probability space.

447. Definition. As introduced above, S is called the *induced transformation* of T on A.

448. Theorem. *Poincaré recurrence theorem. Let $(X, \mathcal{A}, \mu, T)$ be a measure preserving system and suppose $\mu(A) > 0$. Then*

(a) for some $n \in \mathbf{N}$, one has $\mu(A \cap T^{-n}A) > 0$;
(b) $\mu(\{x \in A : A \cap \{Tx, T^2x, T^3x, \ldots\} = \emptyset\}) = 0$.

Idea of proof. For part (a), consider $A, T^{-1}A, T^{-2}A, \ldots, T^{-m}A$ where $m > \frac{1}{\mu(A)}$. These sets cannot be pairwise essentially disjoint. For (b), let B be the set of points in A that don't come back. Apply part (a).   □

449. Theorem. *Let $(X, \mathcal{A}, \mu, T)$ be an ergodic measure-preserving system, and let A be a set of positive measure. Denote by S the induced transformation of T on A. Then $H(S) = \frac{H(T)}{\mu(A)}$.*

Idea of proof. The reader should check that if we know this for $H(T)$ finite, we get it for $H(T) = \infty$ for free. Let then P be a finite generator for T. Define a new (countable) alphabet whose letters are finite strings of letters in P. A point $x \in A$ is assigned a particular string if that string is the P-name of $(x, Tx, T^2x, \ldots, T^{i-1}x)$, where i is the least positive integer with $T^i x \in A$.

450. Exercise. Show that this partition is a generator for S. •

By the Birkhoff ergodic theorem, for a typical $x \in A$ the number of times x returns to A under iteration by T by time n is about $n\mu(A)$. Hence names of size close to $n\mu(A)$ in the induced process look like names of size close to n in the original process.

This would pretty much complete the argument if the partition on A were finite, but as it stands, it proves nothing. We can make it finite by lumping together points in A whose return time is greater than some large number M. This does not seriously alter the entropy. Just be careful in applying Lemma 443.

Sketch of proof. For $x \in A$ let $i(x)$ be the minimum positive integer i such that $T^i x \in A$. By Poincaré, this is $< \infty$ almost everywhere. Put $A_i = \{x : i(x) = i\}$.

451. Exercise. Show that $\sum_{i=1}^{\infty} i\nu(A_i) = \frac{1}{\mu(A)}$. •

Let f be the function taking on the value i on A_i and 0 off A. By the above exercise, $\int f \, d\mu = 1$. For $m \in \mathbb{N}$, let $\overline{P}_m$ be the partition whose letters consist of all P-names of length less than m, plus an additional letter "Z", that assigns to point x the P-name $x_0 x_1 \cdots x_{i(x)-1}$ if $i(x) < m$ and "Z" otherwise. (So the P-name can be read off the $\overline{P}_m$-name except for return times of m and greater, which are lumped into a single cell.) Then $\overline{P} = \bigvee \overline{P}_m$ is a generator for S and hence $H(S) = \lim H(\overline{P}_m, S)$. Choose $\epsilon > 0$ arbitrarily small and pick $m = m(\epsilon)$ such that $\sum_{i=m}^{\infty} i\mu(A_i) < \epsilon$ and such that $H_m = H(\overline{P}_m, S) > H(S) - \epsilon$.

For $N \in \mathbb{N}$ and $x \in A$, let $n(x, N)$ be the time of the Nth return of x to A under iteration by T. Thus it takes a word of length $n(x, N)$ in the (P, T) process to translate to a word of length N in the $(\overline{P}_m, S)$ process. Now choose N large enough that

(1) Most of the space is taken up by reasonable names for the $(\overline{P}_m, S)$ process. That is, for most points $x \in A$, the size of the N-name of x in the $\overline{P}_m$ alphabet is between $2^{-N(H_m+\epsilon)}$ and $2^{-N(H_m-\epsilon)}$.

(2) For most points $x \in A$, $N(\frac{1}{\mu(A)} - \epsilon) < n(x, N) < N(\frac{1}{\mu(A)} + \epsilon)$.

(3) For most points $x \in A$, $\frac{1}{n(x,N)} \sum_{i=0}^{n(x,N)-1} f(T^i x) < \epsilon$.

(4) For most points $x \in A$, the size of the $n(x, N)$-name for P of x is between $2^{-n(x,N)(H+\epsilon)}$ and $2^{-n(x,N)(H-\epsilon)}$. (Here of course $H = H(P, T)$.)

We consider a point "good" if it and most of the points in its names (P-name of length $n(x, N)$ and $\overline{P}_m$-name of length N) satisfy (1) through (4) above.

Now consider a typical "good" point $x \in A$. How big is the $\overline{P}_m$-name of x of length N? According to (1) above, it's at most $2^{-N(H_m-\epsilon)}$. On the other hand, it's at least the size of the P-name of x of length $n(x, N)$, because that entire P-name codes to the $\overline{P}_m$-name of length N. So it's at least $2^{-n(x,N)(H+\epsilon)}$ by 4. This gives $2^{-n(x,N)(H+\epsilon)} \leq 2^{-N(H_m-\epsilon)}$. Taking logs and using (2) above yields $(\frac{1}{\mu(A)} + \epsilon)(H + \epsilon) > (H_m - \epsilon)$. Now letting $\epsilon \to 0$ (so that $m \to \infty$) gives $\frac{1}{\mu(A)} H \geq H(S)$.

Consider again the question, how big the $\overline{P}_m$-name of x of length N is. (More accurately, that part of the name that lies in the good set.) According to (1) above, it's at least $2^{-N(H_m+\epsilon)}$. On the other hand, (3) above says that at most $\epsilon n(x, N)$ characters in the P-name of x of length $n(x, N)$ are part of a block that gets coded to the character "Z". By increasing $|P|$ if necessary, we can assume that which cell of P you are in tells whether or not you are in A. It follows that at most $|P|^{\epsilon N(\frac{1}{\mu(A)}+\epsilon)}$ "good" P-names of various lengths of at least $N(\frac{1}{\mu(A)} - \epsilon)$, and hence of size at most $2^{-N(\frac{1}{\mu(A)}-\epsilon)(H-\epsilon)}$, code to the $\overline{P}_m$-name of x of length N. Hence the $\overline{P}_m$-name of x of length N has size at most $2^{-N(\frac{1}{\mu(A)}-\epsilon)(H-\epsilon)}|P|^{\epsilon N(\frac{1}{\mu(A)}+\epsilon)} \geq 2^{-N(H_m+\epsilon)}$. Taking logs, dividing through by N and letting $\epsilon \to 0$ yields $\frac{1}{\mu(A)} H \leq H(S)$. $\square$

452. Comment. There is another way to prove the above theorem, but in order to indicate it, we need to introduce a generalization of the dbar metric, due to J. Feldman.

453. Definition. The *f-distance* between words $\lambda_1\lambda_2 \cdots \lambda_n$ and $\gamma_1\gamma_2 \cdots \gamma_n$ is $\frac{1}{n}$ times the number of letters that must be removed from each so that the remaining words, when collapsed, are the same.[78] So for example $f(dsornstein, katznelson) = \frac{7}{10}$ since the length of each is ten and a longest common collapsible subword is *ten*.

454. Definition. Let Λ be an alphabet and $n \in \mathbf{N}$. The *fbar distance* $\overline{f}(\mu, \nu)$ between two measures on the words of length n from Λ is the infimum of the expected f-distance between coupled words, taken over all couplings of the two measures.

Let $(X_i)_{i=1}^{\infty}$ and $(Y_i)_{i=1}^{\infty}$ be stationary processes onto a countable alphabet Λ. Let

$$\mu_n(\lambda_1 \cdots \lambda_n) = P(X_1 = \lambda_1, \ldots, X_n = \lambda_n),$$
$$\nu_n(\lambda_1 \cdots \lambda_n) = P(Y_1 = \lambda_1, \ldots, Y_n = \lambda_n);$$

---

[78] Let $w = \lambda_1\lambda_2 \cdots \lambda_n$ and $v = \gamma_1\gamma_2 \cdots \gamma_n$. Pick the maximum k such that there are indices $1 \leq i_1 < i_2 < \ldots < i_k \leq n$ and $1 \leq j_1 < j_2 < \ldots < j_k \leq n$ such that the words $\lambda_{i_1}\lambda_{i_2} \cdots \lambda_{i_k}$ and $\gamma_{j_1}\gamma_{j_2} \cdots \gamma_{j_k}$ coincide. Now $f(w, v) = \frac{n-k}{n}$.

μ_n and ν_n are measures on Λ^n, and we denote their fbar distance by $f(n)$. The *fbar distance* between the processes $(X_i)_{i=1}^{\infty}$ and $(Y_i)_{i=1}^{\infty}$ is $\overline{f}\big((X_i),(Y_i)\big) = \lim_{n\to\infty} f(n)$.

455. Theorem. *The function that takes a process to its entropy is continuous in the fbar metric.*

456. Exercise. Prove the theorem. *Hint: see the proof of Theorem 385.* □

457. Idea of Proof of Theorem 449 using fbar. Recall that the proof was a triviality so long as we could assume that the amount of time it took to return to the set A was bounded from above. Well, it is an easy matter to find a set A' for which this is the case, with $\mu(A \triangle A')$ arbitrarily small. Given that entropy is fbar-continuous, we will be done if we can show that the process induced on A' is fbar close to the process induced on A.

458. Exercise. Prove that the process induced on A' is fbar close to the process induced on A. *Hint: refine the given alphabets on A and A' so they agree on $A \cap A'$. Now just kill the parts of trajectories lying outside $A \cap A'$.* □

5

Bernoulli transformations

459. Definition. Let $0 < p < 1$. A random variable X has the *Bernoulli* $(p, 1-p)$-*distribution* if one has $P(X = 1) = p$ and $P(X = -1) = 1 - p$.

A measure-preserving system $(\Omega, \mathcal{A}, \mu, T)$ is said to be *Bernoulli* if it is isomorphic to an independent process on a finite alphabet.

460. Discussion. The archetypical deep result of isomorphism theory is the Ornstein isomorphism theorem, proved by D. Ornstein in 1970. There are various formulations of this theorem, of which the best known is perhaps that any two Bernoulli systems of the same entropy are isomorphic. A simplified proof of this version of the Ornstein isomorphism theorem was given by M. S. Keane and M. Smorodinsky in 1979. However, Ornstein's original proof is stronger in that it gives a condition, the so-called finite-determination condition, that is on the one hand both necessary and sufficient for a system to be Bernoulli, and on the other hand is easily checked for a number of specific systems.

Indeed, there are several characterizations of the class of Bernoulli systems of independent interest. Experts in isomorphism theory have long known what these characterizations are, and that they are equivalent to each other, though good expositions on many of the equivalences has appeared only recently; Shields (1996) and Thouvenot (2002) are two examples. Indeed, this chapter covers much the same ground as Chapter IV of Shields (1996), with a few differences: Shields' "ABI" (*almost block independent*) processes are essentially the same as our "IC" (*independent concatenation*) processes, and his "ABUP" (*almost blowing up*) property is very similar to what we call "EX" (*extremal*). We both consider the finitely determined (FD) and very weak Bernoulli (VWB) characterizations, and both prove equivalence of four conditions with a circuit of five implications. Our circuit is quite different, however: only two of our five implications appear in Shields (1996). In particular, we do not show directly that FD implies VWB. (This classical result is due to Ornstein and Weiss 1974.)

5.1. The cast of characters

In this subchapter, we introduce the various conditions. For each of these conditions, we will denote the class of stationary processes satisfying the condition by a boldfaced abbreviation. Then we formulate the theorem stating that they coincide (minus one implication, which is handled in the next chapter).

461. Convention. Throughout this chapter and the next, all measure-preserving systems are assumed to be, or be isomorphic to, processes on a finite alphabet. (In particular, we assume that our systems have finite entropy.)

462. Definition. (*Ornstein* 1974.) A stationary process $(X_i)_{i=-\infty}^{\infty}$ is said to be *very weak Bernoulli*, or **VWB**, if the $\bar{d}$-distance between the n-future conditioned on the past and the unconditioned n-future[79] goes to zero in probability as $n \to \infty$.[80]

463. Notation. If $X = (X_i)_{i=-\infty}^{\infty}$ is a stationary process and $n \in \mathbf{N}$, let $\mu_{X,n}$ be the measure X defines on words of length n, that is the measure $\mu_{X,n}(\lambda_1 \cdots \lambda_n) = P(w_i = \lambda_1, 1 \leq i \leq n)$. If X is understood, we will simply use μ_n instead of $\mu_{X,n}$.

464. Definition. (*Ornstein* 1974.) An ergodic stationary process $X = (X_i)_{i=-\infty}^{\infty}$ is said to be *finitely determined*, or **FD**, if for every $\epsilon > 0$ there exist n and $\delta > 0$ such that for any other ergodic process $Y = (Y_i)_{i=-\infty}^{\infty}$ on the same alphabet, if $v(\mu_{X,n}, \mu_{Y,n}) < \delta$ and $|H(X) - H(Y)| < \delta$ then $\bar{d}(X, Y) < \epsilon$.

465. Comment. Thus a process is finitely determined if any process that is a good approximation to it in both distribution and entropy is a good approximation to it in dbar.

466. Definition. Let $(\Omega, \mathcal{A})$ be a measurable space and suppose that μ and ν are measures on $\mathcal{A}$. Let $s = s\left(\frac{\nu}{\mu}\right) = \sup\left\{\frac{\nu(A)}{\mu(A)} : A \in \mathcal{A}\right\}$. If $s < \infty$, we say that ν is a *submeasure of μ having size* $\frac{1}{s}$.

467. Definition. (*J. Thouvenot* 2002.) A (not necessarily stationary) process μ is called *extremal* if for every ϵ, there is a δ, such that for all sufficiently large n, there is a set of measure less than δ (that we call "error") of words of length n, such that the following holds: Any submeasure of μ_n of size greater than $2^{-\delta n}$ whose support is off error is within ϵ of μ_n in dbar. If one can always take the error set to be empty, the process is *completely extremal*.[81]

---

[79] This is a function from pasts to non-negative reals.

[80] To clarify: if, for every $\epsilon > 0$, there exists some N such that, for every fixed $n > N$, there is a *past*-measurable set B with $P(B) < \epsilon$ such that for every $(x_i)_{i=-\infty}^{\infty} \in B^c$, if we define measures μ_n and ν_n on words of length n by $\mu_n(\lambda_1\lambda_2 \cdots \lambda_n) = P(w_i = \lambda_i, 1 \leq i \leq n)$ and $\nu_n(\lambda_1\lambda_2 \cdots \lambda_n) = P(w_i = \lambda_i, 1 \leq i \leq n | w_i = x_i, i \leq 0)$, then $\bar{d}(\mu_n, \nu_n) < \epsilon$.

[81] Here is a more precise definition. A process (possibly not stationary) $(X_i)_{i=-\infty}^{\infty}$ is said to be extremal if for every $\epsilon > 0$ there exists $\delta > 0$ and $N \in \mathbf{N}$ such that for every $n \geq N$ there is a

We now introduce some terminology that we feel is useful in conceptualizing submeasures, extremality and related notions.

468. Definition. Let Λ be a finite alphabet, let μ be a measure on $\Lambda^{\mathbf{Z}}$ and for each $n \in \mathbf{N}$, let ν_n be a measure on Λ^n. We say that the sequence (ν_n) is *exponentially fat* relative to μ if $\lim_n \frac{1}{n} \log s \left(\frac{\nu_n}{\mu_n} \right) = 0$.

469. Abbreviated definition of extremal. A process is extremal if there is a sequence of error sets having measure tending to zero such that for any exponentially fat sequence (ν_n) of submeasures of μ_n whose supports are off the error sets, one has $\lim_n \overline{d}(\mu_n, \nu_n) = 0$.

470. Further abbreviated definition of extremal. A process is extremal if for sufficiently large n, exponentially fat submeasures whose support is off some small subset are dbar close to the whole measure.

471. Definition. We denote by **EX** the class of all extremal processes that are also stationary.

472. Important comment. In practice we will generally, in what is admittedly an abuse of terminology, use "exponentially fat" to refer not to a sequence of measures but to a single measure ν_n on words of length n, where n is large. Basically, we mean by this that $s(\frac{\nu_n}{\mu_n}) \leq 2^{\delta n}$ for some $\delta > 0$ fixed in advance. Hence we can further abbreviate the definition of extremal thus: *A process is extremal if for sufficiently large n, exponentially fat submeasures whose support is off a small error set must be dbar close to μ_n.*

473. Definition. A process $(X_i)_{i=-\infty}^{\infty}$ is an *independent concatenation* if it is obtained by taking any measure on words of length n, using that measure to choose the first n letters, using that measure again to choose the next n letters independently, and so on forever (and backwards in time).[82]

474. Definition. If, in the definition of independent concatenation, the original measure on words of length n comes from a given process $(Y_i)_{i=-\infty}^{\infty}$, then the independent concatenation $(X_i)_{i=-\infty}^{\infty}$ will be called the *period n independent concatenation induced by* $(Y_i)_{i=-\infty}^{\infty}$.

---

set $B_n \subset \Lambda^n$ with $\mu_n(B_n) < \epsilon$ having the property that for any submeasure ν of μ_n of size at least $2^{-\delta n}$ (that is, with $s \left(\frac{\nu}{\mu_n} \right) \leq 2^{\delta n}$) satisfying $\nu(B_n) = 0, \overline{d}(\mu_n, \nu) \leq \epsilon$. If one can always take $B_n = \emptyset$, $(X_i)_{i=-\infty}^{\infty}$ is *completely extremal*.

[82] Thus, if there is an $n \in \mathbf{N}$ and a measure μ on words of length n such that, letting, for $k \in \mathbf{Z}$ and $\lambda = \lambda_1 \cdots \lambda_n$ a word of length n, $E_{k,\lambda}$ be the event $\left\{ (w_i)_{i=-\infty}^{\infty} : w_{kn+i} = \lambda_i, 1 \leq i \leq n \right\}$, one has that the probability of $E_{k,\lambda}$ is $\mu(\lambda)$ and any collection of events $\left\{ E_{k_i, \lambda^{(i)}} : 1 \leq i \leq r \right\}$, for distinct $k_1, \cdots, k_r$, is independent.

475. Comment. Independent concatenations need not be stationary.

The following definition is equivalent to P. Shields' *almost block indepen-dence*. (See Shields 1979, in which proves this property to be equivalent to Bernoulli.)

476. Definition. A stationary process $Y = (Y_i)_{i=-\infty}^{\infty}$ is *a dbar limit of independent concatenations*, or **IC**, if it is the dbar limit of a sequence of independent concatenations, i.e. if there exists a sequence $X^{(n)}$ of independent concatenations such that $\lim_{n\to\infty} \bar{d}\left(Y, X^{(n)}\right) = 0$.

477. Comment. The reader should note that it follows immediately from the definition that **IC** is closed under passage to dbar limits.

478. Notation. We will use **B** to denote the class of Bernoulli processes and **FB** to denote the class of their factors.

The boldfaced theorem.

The main purpose of this chapter is to establish the following.

479. Theorem. $\boldsymbol{B \subset FB \subset IC = EX = VWB = FD}$.

480. Comment. The rest of this chapter constitutes a proof of the boldfaced theorem. This will be accomplished in a series of steps. We'll show **FD** ⊂ **B** in the next chapter, thus completing the proof that all of the boldfaced properties are equivalent. The first step we can do immediately, the others will be done in the subchapters to come.

Step 1: B ⊂ FB

This is obvious. Any process is a factor of itself.

5.2. Step 2: FB ⊂ IC

Step 2 will be accomplished in two stages. First, we prepare an explication of Definition 442.

Let $(X_i)_{i=-\infty}^{\infty}$ be a stationary process on a finite alphabet Λ, let Λ' be another finite alphabet, and suppose that there are given some $m \in \mathbf{N}$ and a function $f : \Lambda^{\{-m,-m+1,\ldots,m\}} \to \Lambda'$. We define a process $(Y_i)_{i=-\infty}^{\infty}$ by $Y_i = f(X_{i-m}, X_{i-m+1}, \ldots, X_{i+m})$.

481. Exercise. Show that $(Y_i)_{i=-\infty}^{\infty}$ is a stationary process, and that moreover it is a factor of $(X_i)_{i=-\infty}^{\infty}$. *Hint: try $\pi(x) = y$ as factor map, where $y_i = f(x_{i-m}, x_{i-m+1}, \ldots, x_{i+m})$.*

The process $(Y_i)_{i=-\infty}^{\infty}$ constructed above is a coding of the process $(X_i)_{i=-\infty}^{\infty}$. The function f is the coding map, or code. Thus we see that the codings of $(X_i)_{i=-\infty}^{\infty}$ are a subclass of the factors of $(Y_i)_{i=-\infty}^{\infty}$. Our strategy is as follows. First, we will show that any factor of a stationary process on a finite alphabet can be dbar-approximated to any desired degree of accuracy by a coding of that process. Then, we will show that any coding of a Bernoulli process is **IC**.

482. Exercise. Verify that the above scheme is sufficient for completing Step 2.

483. Theorem. *Let $(X_i)_{i=-\infty}^{\infty}$ be a stationary process on a finite alphabet Λ and let $(Z_i)_{i=-\infty}^{\infty}$, a stationary process on a finite alphabet Λ', be one of its factors.[83] For every $\epsilon > 0$, there exists a coding $(Y_i)_{i=-\infty}^{\infty}$ of $(X_i)_{i=-\infty}^{\infty}$ such that $\overline{d}\big((Z_i)_{i=-\infty}^{\infty}, (Y_i)_{i=-\infty}^{\infty}\big) < \epsilon$.*

Idea of proof. Let π be the factor map which takes doubly infinite strings $(x_i)_{i=-\infty}^{\infty}$ to strings $(z_i)_{i=-\infty}^{\infty}$. For each $\lambda' \in \Lambda'$, put $E_{\lambda'} = \big\{(z_i)_{i=-\infty}^{\infty} : z_0 = \lambda'\big\}$.

484. Exercise. For some sufficiently large m, there is a partition $\{D_{\lambda'} : \lambda' \in \Lambda'\}$ of $\Lambda^{\mathbb{Z}}$ such that $\sum_{\lambda' \in \Lambda'} P\big(\pi^{-1}(E_{\lambda'}) \triangle D_{\lambda'}\big) < \epsilon$ and such that each $D_{\lambda'}$ is a union of cylinder sets having support in $\{-m, -m+1, \ldots, m\}$. •

We now define the coding map $f : \Lambda^{\{-m, -m+1, \ldots, m\}} \to \Lambda'$. The previous exercise implies that for each located word w on the alphabet Λ having support in $\{-m, -m+1, \ldots, m\}$, $\varphi(w)$ lies entirely inside some $D_{\lambda'}$; we let $f(w) = \lambda'$ and form the associated coding $(Y_i)_{i=-\infty}^{\infty}$. Now $(Y_i)_{i=-\infty}^{\infty}$ is ϵ-close to $(Z_i)_{i=-\infty}^{\infty}$ in variation, which is of course even stronger than being ϵ-close in dbar. □

Now for the remaining piece.

485. Theorem. *Let $(X_i)_{i=-\infty}^{\infty}$ be a Bernoulli process and let $(Y_i)_{i=-\infty}^{\infty}$ be a coding of $(X_i)_{i=-\infty}^{\infty}$. Then $(Y_i)_{i=-\infty}^{\infty}$ is **IC**.*

Sketch of proof. Let m and $f : \Lambda^{\{-m, -m+1, \ldots, m\}} \to \Lambda'$ be as in the coding construction introduced above. Let $\epsilon > 0$ be arbitrary. Choose $n > \frac{m}{2\epsilon}$ and let $(Z_i)_{i=-\infty}^{\infty}$ be the period n independent concatenation induced by $(Y_i)_{i=-\infty}^{\infty}$. We couple $(Z_i)_{i=-\infty}^{\infty}$ and $(Y_i)_{i=-\infty}^{\infty}$ by induction. We couple blocks of times $\{kn+1, kn+2, \ldots, kn+n\}$ in turn for $k = 0, 1, 2, 3, \ldots$ Inside the blocks, we will take care to use the diagonal coupling on the conditionals often enough to

---

[83] It is a consequence of the Krieger finite generator theorem that every factor may be so represented. Indeed, one may take $\Lambda = \Lambda'$, if desired.

ensure a low dbar distance. We will be able to do this when the probabilities for the various letters at the next time conditioned on previously coupled times are equal for the two processes. This is unlikely to be the case at the beginning of a block; Y_{kn+1} will depend crucially on $Y_{kn-m+1}, Y_{kn-m+2}, \ldots, Y_{kn}$, while Z_{kn+1} will not depend on $Z_{kn-m+1}, Z_{kn-m+2}, \ldots, Z_{kn}$ (or any other previously coupled time). However, since Y_i and Y_j are independent when $|i - j| > 2m$, prospects seem good on the interior of the blocks. Be warned, however, that there is some subtlety at play here; taking $m = 1$ and $n = 10$, while it's true that, for example, Y_{13} and Y_{10} must be independent, it need not be the case that $P(Y_{13} = b|Y_{12} = Y_{11} = a) = P(Y_{13} = b|Y_{12} = Y_{11} = Y_{10} = a)$. (Whereas of course $P(Z_{13} = b|Z_{12} = Z_{11} = a) = P(Z_{13} = b|Z_{12} = Z_{11} = Z_{10} = a)$.)

The easiest way to solve this problem is to skip a sub-block of size $2m$ at the beginning of each n-block. That is, we couple each n-block in the order $\{nk + 2m + 1, nk + 2m2, \ldots, nk + n, nk + 1, nk + 2, \ldots, nk + 2m\}$. We couple the conditionals with the diagonal measure whenever possible, which will be the case for at least $n - 2m$ of the n times in the block (the exceptional times are the first $2m$ times).

486. Exercise. Fill in the details and complete the proof. $\qquad\qquad\square$

The proof of this step is now complete, however, we pause to formulate the following convenient corollary. Together with the obvious fact that **IC** is closed under dbar limits, it shows that **IC** is closed under the taking of factors.

487. Corollary. *IC is closed under finite code factors.*

Idea of proof. Let (X_i) be **IC** and let (Y_i) be a finite coding of (X_i). Let M be the length of the code. Since (X_i) is IC we can approximate it much closer than $\frac{1}{M}$ in dbar by an independent concatenation (Z_i). Let $n \gg M$ be a period of the independent concatenation. Since (Z_i) is dbar close to (X_i) we can assume that they are coupled in such a way that most of their M-words coincide. Hence, if we let (W_i) be the coding of (Z_i) that comes from the exact same code used to derive (Y_i) from (X_i), (W_i) will be close to (Y_i) in dbar. On the other hand, if we only use the code on the interior of the n-blocks, and begin and end each n-block with some arbitrary fixed word of length M, and call the result of this (U_i), then (U_i) will be an independent concatenation and will be close to (W_i) in dbar, hence close to (Y_i) in dbar as well. $\qquad\qquad\square$

5.3. Step 3: IC ⊂ EX

Our first task is to show that independent processes are completely extremal, which will require some preparation. After doing that, we will argue briefly that independent concatenations are extremal for much the same reasons and finally that the class of extremal processes is closed under dbar limits, which will complete Step 3.

Let $(X_i)_{i=-\infty}^{\infty}$ be a stationary process on a countable alphabet Λ and let $(\Omega, \mathcal{A}, \mu, T)$ be the standard realization of the process as a measure-preserving system. ($\Omega = \Lambda^{\mathbf{Z}}$, etc.) For $n \in \mathbf{N}$, let μ_n be the measure on Λ^n defined by $\mu_n(w) = \mu(\varphi(w))$.

488. Comment. We formulate the following theorem two times, first for the "Idea" section, then more formally for the more detailed proof sketch.

489. Theorem. *Let (Ω, μ) be a probability space and let X_i from Ω to $\{-1, 1\}$ be independent random variables each assigning probability $\frac{1}{2}$ to both -1 and 1. Let $S_n = \sum_{i=1}^{n} X_i$. Let ν be a submeasure of μ and let E be the expectation operator corresponding to ν. For every $\epsilon > 0$ there is a $\delta > 0$ such that if n is sufficiently large and $E(S_n) > \epsilon n$ then the size of ν is not more than $2^{-\delta n}$. (In other words, ν is not exponentially fat.)*

Idea of proof. On the one hand, $E(S_n) > \epsilon n$. On the other hand, if R_n is the set of names in $\{-1, 1\}^n$ such that $S_n > \frac{\epsilon n}{2}$, then by an elementary fact about random walks, (R_n) is not exponentially fat. From these facts, the reader can establish that ν is not exponentially fat.

490. Reformation of Theorem 489. *Let $\epsilon > 0$. There exists a $\delta > 0$ and an $N \in \mathbf{N}$ such that for every $n \geq N$, if $(\Omega, \mathcal{A}, \mu)$ is a probability space and $X_1, \ldots, X_n$ are independent $\left(\frac{1}{2}, \frac{1}{2}\right)$-Bernoulli random variables, then for any measure ν on $\mathcal{A}$, if $E_\nu \left(\sum_{i=1}^{n} X_i\right) > \epsilon n$, where E_ν denotes expectation relative to the measure ν, then $s(\frac{\nu}{\mu}) > 2^{\delta n}$.*

Sketch of proof. Notice first that $E_\nu \left(\sum_{i=1}^{n} X_i\right) > \epsilon n$ implies $P_\nu \left(\sum_{i=1}^{n} X_i > \frac{\epsilon}{2} n\right) > \frac{\epsilon}{2n}$. Therefore we will be done if we exhibit $\delta > 0$ such that for all relevant $(\Omega, \mathcal{A}, \mu)$ and (X_i) and all sufficiently large n, $P_\mu \left(\sum_{i=1}^{n} X_i > \frac{\epsilon}{2} n\right) < 2^{-\delta n} \frac{\epsilon}{2n}$.

Let $f(x) = \frac{1}{\left(\frac{1}{2} + \frac{x}{4}\right)\left(\frac{1}{2} - \frac{x}{4}\right)}$.

491. Exercise. Show that f is strictly decreasing on $[0, 1)$.   ●

Note that f has a maximum value of 4 at 0. Choose $\delta > 0$ so small that $2^{-\delta} > \sqrt{\frac{f(\epsilon)}{4}}$.

For large n, let $j = j(n) = \left\lceil \left(\frac{1}{2} + \frac{\epsilon}{4}\right) n \right\rceil$. Then

$$P_\mu\left(\sum_{i=1}^{n} X_i > \frac{\epsilon}{2} n\right) = 2^{-n} \sum_{i=j}^{n} \binom{n}{i} \leq 2^{-n} \left(\frac{1}{2} n\right) \binom{n}{j}$$

$$\approx \frac{n^n e^{-n} \sqrt{2\pi n}}{j^j e^{-j} \sqrt{2\pi j}(n-j)^{n-j} e^{-(n-j)} \sqrt{2\pi(n-j)}}$$

$$= \left(\frac{n}{j}\right)^j \left(\frac{n}{n-j}\right)^{n-j} \sqrt{\frac{n}{2\pi j(n-j)}} (2^{-n}) \left(\frac{1}{2}n\right).$$

492. Exercise. Show that for large n, this is less than $\left(\frac{n^2}{j(n-j)}\right)^{\frac{1}{2}n} 2^{-n} \frac{\epsilon}{2n}$. *Hint:*
$\frac{n-j}{j} < 1 - \epsilon$. *For large n,* $(1-\epsilon)^{\frac{1}{4}\epsilon n} \left(\frac{n^2}{\epsilon}\right) \ll 1$.   ●

Therefore we have, for all sufficiently large n,

$$P_\mu\left(\sum_{i=1}^{n} X_i > \frac{\epsilon}{2} n\right) \leq \left(\frac{n^2}{j(n-j)}\right)^{\frac{1}{2}n} 2^{-n} \frac{\epsilon}{2n}$$

$$\leq \left(\frac{1}{\left(\frac{1}{2} + \frac{\epsilon}{4}\right)\left(\frac{1}{2} - \frac{\epsilon}{4}\right)}\right)^{\frac{1}{2}n} 2^{-n} \frac{\epsilon}{2n} = \left(\frac{f(\epsilon)}{4}\right)^{\frac{1}{2}n} \left(\frac{\epsilon}{2n}\right)$$

$$\leq 2^{-\delta n} \left(\frac{\epsilon}{2n}\right).$$ □

493. Exercise. Let $\epsilon > 0$, let P be a finite set and suppose μ and ν are two probability measures on P having variation distance $v(\mu, \nu) \geq \epsilon$. There exists a set $S \subset P \times [0, 1)$ such that $\mu \times m(S) = \frac{1}{2}$ and $\nu \times m(S) \geq \frac{1}{2} + \frac{\epsilon}{2}$. *Hint: partition P into two cells P_1, P_2 with $\mu \geq \nu$ on P_1 and $\nu > \mu$ on P_2. If $\mu(P_2) \geq \frac{1}{2}$, take $S = P_2 \times [0, \frac{1}{2\mu(P_2)})$.*

494. Definition. For X a finite set, μ a measure on X and S a subset of X, we write $\mu^{(S)}(A) = \frac{\mu(S \cap A)}{\mu(S)}$. $\mu^{(S)}$ is called the *normalized probability measure induced by μ on S.*

495. Exercise. Let Λ be a finite alphabet, let $0 < t < 1$, $n \in \mathbf{N}$ and suppose that μ, ν_1, ν_2 and ν are probability measures on Λ^n satisfying $\nu = t\nu_1 + (1 - t)\nu_2$. Show that $\bar{d}(\mu, \nu) \leq t\bar{d}(\mu, \nu_1) + (1-t)\bar{d}(\mu, \nu_2)$. *Hint: if c_i is a coupling of μ with ν_i, $i = 1, 2$, then $c = tc_1 + (1 - t)c_2$ is a coupling of μ and ν.*

496. Exercise. More generally, if $\nu = \sum_{i=1}^{t} a_i \nu_i$, where $\sum_{i=1}^{t} a_i = 1$ and $a_i > 0$, $1 \leq i \leq t$, then $\bar{d}(\mu, \nu) \leq \sum_{i=1}^{t} a_i \bar{d}(\mu, \nu_i)$. Also, one has $s\left(\frac{\nu}{\mu}\right) \leq \max_i s\left(\frac{\nu_i}{\mu}\right)$.

497. Exercise. Let Λ be a finite alphabet, let $n \in \mathbf{N}$ and suppose that μ and ν are measures on Λ^n. Show that $v(\mu, \nu) \geq \bar{d}(\mu, \nu)$.

498. Theorem. *Let X be a finite set, let μ and ν be measures on X with $s \geq s(\frac{\nu}{\mu})$. Let $\beta = \max_{x \in X} \mu(x)$. Let $\gamma > 0$ with $1 - \gamma < \frac{1}{1+\beta s}$. Then there exist t and a collection of subsets $S_i \subset X$ such that $\mu(S_i) \geq \frac{1}{s}$, $1 \leq i \leq t$, and $\nu(\nu, \sum_{i=1}^t \frac{1}{t} \mu^{(S_i)}) < \gamma$.*

499. Comment. Recall that $\nu(\nu, \sum_{i=1}^t \mu^{(S_i)}) = \frac{1}{2} \sum_{x \in X} |\nu(x) - \frac{1}{t} \sum_{i=1}^t \mu^{(S_i)}(x)|$.

Idea of proof. Pick your sets S_i with $\frac{1}{s} \leq \mu(S_i) \leq \frac{1}{s} + \beta$ and such that any point x is in a fraction of the sets close to $f(x) = \frac{\nu(x)}{s\mu(x)}$. This can be done in pretty much any ad hoc way. At any step, say in the construction of S_i, having previously constructed $S_1, \ldots, S_{i-1}$, go through all the points $x \in X$ and throw them in S_i if, up to that point, they have been in too small a proportion of the sets. If you run out of these, throw points in starting at the beginning of the list. In any event, stop as soon as your set gets as large as $\frac{1}{s}$ in measure. As the number n of sets gets large, all of the points will have been in about the right fraction of the sets, which is sufficient for the result. □

500. Discussion. Suppose that you wake up one morning having the insane desire to couple the northernmost half of the Earth to the whole of the Earth, in such a way that the average distance between coupled points is less than the length of your foot. It is rather obvious that (unless you have really big feet) your desire must go unfulfilled: it's possible for half of the Earth to be inextricably "far from" the whole. But, as our next theorem will show, the same isn't true for $\{0, 1\}^n$ with mean Hamming distance. Not only is it impossible to extract half the space such that the half you extract is "far away from" the whole space when n is large; it is even impossible to extract an exponentially fat subset such that the subset is far away from the whole space. In other words, $\{0, 1\}^n$ is completely extremal.

501. Theorem. *An independent process on a finite alphabet is completely extremal.*

502. Comment. As we often do in this book, we will give the proof in two different grades of detail, however in this case there is a bit more repetition than usual and a few notational differences. As always, we strongly suggest that readers attempt to work through the "Idea" section on their own before consulting the sketch.

503. First reformulation of Theorem 501. *Let $P = \{p_1, p_2, p_3, \ldots, p_m\}$ be a finite set and let μ be a measure on P. For an understood $n \in \mathbf{N}$, let $\Omega = P^n$ and let Φ denote the measure μ^n on Ω. Let $S \subset \Omega$ and let $\varphi = \mu^{(S)}$ be the normalized probability measure on S induced by Φ. We will show that for* *any c, there is a d such that if n is large enough and* $\overline{d}\big((S, \varphi), (\Omega, \Phi)\big) > c$ *then* $\varphi(S) < 2^{-dn}$. *(What this all says is that* (P^∞, μ^∞) *is completely extremal.)*

Idea of proof. Let I be the unit interval and let m be Lebesgue measure on I. We consider the space $\Omega \times I^n$ with measure $\Phi \times m^n$. Similarly, we will look at the space $S \times I^n$ with measure $\varphi \times m^n$. Our goal is to prove that $\Phi(S) < 2^{-dn}$, which is equivalent to $(\Phi \times m^n)(S \times I^n) < 2^{-dn}$.

Our first step is to construct independent random variables on $(\Omega \times I^n, \Phi \times m^n)$ such that each has probability $\frac{1}{2}$ of being 1 and $\frac{1}{2}$ of being -1. Let $k < n$ and select some word $a_1 a_2 a_3 \cdots a_k$. (If $k = 0$ we mean the empty word.) If $a_1 a_2 a_3 \cdots a_k$ occurs as the initial word of some member of S (as is always true when $k = 0$ if $S \neq \emptyset$) we write $a_1 a_2 a_3 \cdots a_k \in i(S)$ and let $v_{a_1 a_2 a_3 \cdots a_k}$ be the conditional measure (under φ) of the $(k + 1)$st letter of a word in S given that the first k letters are $a_1 a_2 a_3 \cdots a_k$. For such a word, let $d_{a_1 a_2 a_3 \cdots a_k} = v(\mu, v_{a_1 a_2 a_3 \cdots a_k})$, where v is variation distance. Invoking Lemma 493, let $\theta = \theta(a_1 a_2 a_3 \cdots a_k)$ be a subset of $P \times I$ such that $(\mu \times m)(\theta) = \frac{1}{2}$ and $(v_{a_1 a_2 a_3 \cdots a_k} \times m)(\theta) > \frac{1}{2} + \frac{d_{a_1 a_2 a_3 \cdots a_k}}{4}$.

Now we define a random variable X_{k+1} on $\Omega \times I^n$:

$$
X_{k+1}\big((a_1 a_2 \cdots a_n), (t_1 t_2 \cdots t_n)\big)
$$
$$
= \begin{cases}
-1 & \text{if} \quad a_1 a_2 \cdots a_k \notin i(S) \text{ and } t_{k+1} \leq \frac{1}{2} \\
1 & \text{if} \quad a_1 a_2 \cdots a_k \notin i(S) \text{ and } t_{k+1} > \frac{1}{2} \\
1 & \text{if} \quad a_1 a_2 \cdots a_k \in i(S) \text{ and } (a_{k+1}, t_{k+1}) \in \theta_{a_1 a_2 \cdots a_k} \\
-1 & \text{if} \quad a_1 a_2 \cdots a_k \in i(S) \text{ and } (a_{k+1}, t_{k+1}) \notin \theta_{a_1 a_2 \cdots a_k}.
\end{cases}
$$

Under $\Phi \times m^n$, $X_1, X_2, \ldots$ are independent random variables taking on the values $-1, 1$ with probabilities $\frac{1}{2}$ each.

It is time to make use of the fact that $\overline{d}\big((S, \varphi), (\Omega, \Phi)\big) > c$. We couple (S, φ) and (Ω, Φ) by induction. Assume that we already know how the first k coordinates of the elements of S are coupled with the first k coordinates of the elements of Ω. Consider such a coupled pair of k-tuples $(a_1 a_2 a_3 \cdots a_k, b_1 b_2 b_3 \cdots b_k)$; conditioned on that coupled pair, we want the joint distribution of (a_{k+1}, b_{k+1}). a_{k+1} has conditioned distribution $v_{a_1 a_2 a_3 \ldots a_k}$ and b_{k+1} has conditioned distribution μ. Couple those two distributions as closely as possible; the probability that the coordinates don't match is $d_{a_1 a_2 a_3 \ldots a_k}$. The expected mean Hamming distance between words forming a coupled pair for this coupling must be greater than c. That is,

$$
c < \frac{h_1 + h_2 + h_3 + \cdots + h_n}{n},
$$

where h_k is the probability that the kth coordinates differ.

The rest is definition chasing. Letting E be the expectation operator of the measure $\varphi \times m^n$, show that

$$\frac{c}{2} < \frac{1}{n} E \left(\sum_{i=1}^{n} X_i \right)$$

and use Theorem 489.

504. Abbreviation. S far away in dbar, independent $\frac{1}{2}$-$\frac{1}{2}$ (X_i) defined inductively which prejudices for 1 as much as possible on S. Far dbar implies prejudice is strong enough to force partial sums of X_i to grow linearly on S in expectation, proving S shrinks exponentially.

That's it for the idea of the proof. We now give a second reformulation of the theorem and proceed to the proof sketch.

505. *Second reformulation of Theorem 501.* *We formulate a precise statement of what is to be proved. Let $\Lambda = \{\lambda_1, \lambda_2, \ldots, \lambda_m\}$ be a finite alphabet and let p be a measure on Λ. For fixed n, we define a measure μ_n on Λ^n by $\mu_n(\{a_1 a_2 \cdots a_n\}) = \prod_{i=1}^{n} p(\{a_i\})$. Given $\epsilon > 0$, there exists $\delta = \delta(\epsilon) > 0$ such that for all sufficiently large n, if ν is any measure on Λ^n with $\overline{d}(\mu_n, \nu) > \epsilon$ then $s\left(\frac{\nu}{\mu_n}\right) > 2^{\delta n}$.*

Sketch of proof. First we show that it is sufficient to restrict attention to ν of the form μ_{S_i}, where S_i is a subset of Λ^n. That is, it is sufficient to show that:

506. Claim. If $\epsilon > 0$ then for some $\delta > 0$, for all sufficiently large n, if S is any subset of Λ^n with $\overline{d}(\mu_n, \mu_n^{(S)}) > \epsilon$ then $\mu_n(S) \leq 2^{-\delta n}$.

To see that this suffices for the general case, let $\epsilon > 0$ and let δ be chosen such that for all sufficiently large n, if S is any subset of Λ^n with $\overline{d}(\mu_n, \mu_n^{(S)}) > \frac{\epsilon}{2}$ then $\mu_n(S) < 2^{-\delta n}$. We can certainly choose a smaller δ if we want to, so we shall assume without loss of generality that $2^\delta p(\lambda_i) < 1$, $1 \leq i \leq m$. Now suppose that n is large enough that $\frac{1}{1+\beta s} > 1 - \frac{\epsilon}{2}$, where $\beta = \max_{1 \leq i \leq m} p(\lambda_i)^n$ (notice that β is the maximum size of $\mu_n(w)$ for a word $w \in \Lambda^n$) and $s = 2^{\delta n}$. Let ν be any measure on Λ^n such that $s\left(\frac{\nu}{\mu_n}\right) \leq 2^{\delta n} = s$. By Theorem 498, there exist t and a collection of subsets $S_i \subset \Lambda^n$ such that $\mu_n(S_i) \geq 2^{-\delta n}$, $1 \leq i \leq t$, and $\nu(\nu, \frac{1}{t} \sum_{i=1}^{t} \mu_n^{(S_i)}) < \frac{\epsilon}{2}$. By Exercise 497, $\overline{d}(\nu, \frac{1}{t} \sum_{i=1}^{t} \mu_n^{(S_i)}) < \frac{\epsilon}{2}$.

Since $\mu_n(S_i) \geq 2^{-\delta n}$, one has $\overline{d}(\mu_n, \mu_n^{(S_i)}) \leq \frac{\epsilon}{2}$, $1 \leq i \leq t$. This implies by Exercise 496 that

$$\overline{d}(\mu_n, \nu) \leq \overline{d} \left(\mu_n, \frac{1}{t} \sum_{i=1}^{t} \mu_n^{(S_i)} \right) + \overline{d} \left(\frac{1}{t} \sum_{i=1}^{t} \mu_n^{(S_i)}, \nu \right)$$

$$\leq \frac{1}{t} \sum_{i=1}^{t} \overline{d} \left(\mu_n, \mu_n^{(S_i)} \right) + \frac{\epsilon}{2} \leq \epsilon.$$

507. Proof of Claim 506. Let $\epsilon > 0$. Choose δ and N as in Theorem 489. Suppose now that $n > N$ and that $S \subset \Lambda^n$ with $\overline{d}(\mu_n, \mu_n^{(S)}) > \epsilon$. We must show that $\mu_n(S) \leq 2^{-\delta n}$. According to the conclusion of Theorem 489, it is sufficient to construct a probability space $(\Omega, \mathcal{A}, \tilde{\mu})$, a measurable set $S' \in \mathcal{A}$ with $\tilde{\mu}(S') = \mu_n(S)$, and independent $\left(\frac{1}{2}, \frac{1}{2}\right)$-Bernoulli random variables $X_1, \ldots, X_n$ defined on $(\Omega, \mathcal{A}, \tilde{\mu})$ with $E_{\tilde{\mu}}\left(\sum_{i=1}^n X_i(x) | x \in S'\right) > \epsilon n$.

508. Notation. Write $w \in i(S)$ if $w = a_1 a_2 \cdots a_k$ is the initial string of some word in S.

Let $\Omega = \Lambda^n \times [0, 1)^n$, endowed with the measure $\mu = \mu_n \times m^n$, and let $S' = S \times [0, 1)^n$. Plainly $\mu(S') = \mu_n(S)$. For $0 \leq k < n$ and $a_1, \ldots, a_{k+1} \in \Lambda$, if $a_1 a_2 \cdots a_k \in i(S)$ then write

$$v_{a_1 a_2 \cdots a_k}(a_{k+1}) = P_{\mu^n}(w_{k+1} = a_{k+1} | w_i = a_i, 1 \leq i \leq k, w_1 \cdots w_n \in S).$$

(When $k = 0$, the string $a_1 a_2 \cdots a_k$ is of course empty; we denote it by $\emptyset$ and consider it to be a member of $i(S)$, writing $v_\emptyset$, etc.) Now write $d_{a_1 a_2 \cdots a_k} = v(\mu, v_{a_1 a_2 \cdots a_k})$, and choose by Exercise 493 a set $\theta_{a_1 a_2 \cdots a_k} \subset \Lambda \times [0, 1)$ with $\mu \times m(\theta_{a_1 a_2 \cdots a_k}) = \frac{1}{2}$ and $v_{a_1 a_2 \cdots a_k} \times m(\theta_{a_1 a_2 \cdots a_k}) \geq \frac{1}{2} + \frac{1}{2} d_{a_1 a_2 \cdots a_k}$. Now for $0 \leq k < n$ put

$$X_{k+1}\left((a_1 a_2 \cdots a_n), (t_1 t_2 \cdots t_n)\right)$$

$$= \begin{cases} -1 & \text{if } a_1 a_2 \cdots a_k \notin i(S) \text{ and } t_{k+1} \leq \frac{1}{2} \\ 1 & \text{if } a_1 a_2 \cdots a_k \notin i(S) \text{ and } t_{k+1} > \frac{1}{2} \\ 1 & \text{if } a_1 a_2 \cdots a_k \in i(S) \text{ and } (a_{k+1}, t_{k+1}) \in \theta_{a_1 a_2 \cdots a_k} \\ -1 & \text{if } a_1 a_2 \cdots a_k \in i(S) \text{ and } (a_{k+1}, t_{k+1}) \notin \theta_{a_1 a_2 \cdots a_k}. \end{cases}$$

Clearly the X_i are $\left(\frac{1}{2}, \frac{1}{2}\right)$-Bernoulli random variables. We must show they are independent. We do a special case of this here to give the general idea and leave the general case as an exercise. We claim that $P(X_2 = 1, X_3 = -1) = \frac{1}{4}$. First note that $X_2 = 1$ and $X_3 = -1$ when evaluated at $\left((a_1 a_2 \cdots a_n), (t_1 t_2 \cdots t_n)\right)$ in three cases. Case 1: $a_1 \notin i(S)$, $t_2 > \frac{1}{2}$, $t_3 \leq \frac{1}{2}$. Case 2: $a_1 \in i(S)$, $(a_2, t_2) \in \theta_{a_1}$, $a_1 a_2 \in i(S)$, $(a_3, t_3) \notin \theta_{a_1 a_2}$. Case 3: $a_1 \in i(S)$, $(a_2, t_2) \in \theta_{a_1}$, $a_1 a_2 \notin i(S)$, $t_3 \leq \frac{1}{2}$. Letting $p = P(a_1 \notin i(S))$, and letting $q = P(a_1 a_2 \notin i(S) | a_1 \in i(S))$, then summing the probabilities of these three cases we get $p \cdot \frac{1}{2} \cdot \frac{1}{2} + (1 - p) \cdot \frac{1}{2} \cdot (1 - q) \cdot \frac{1}{2} + (1 - p) \cdot \frac{1}{2} \cdot q \cdot \frac{1}{2} = \frac{1}{4}$.

509. Exercise. Do the general case. That is, establish that $X_1, \ldots, X_n$ are independent.   •

Our next observation is that for a fixed word $w_1 \cdots w_k \in i(S)$, one has that

$$E\left(X_{k+1}(a, t) | a_1 \cdots a_n \in S, a_i = w_i, 1 \leq i \leq k\right)$$
$$= v_{w_1 w_2 \cdots w_k} \times m(\theta_{w_1 w_2 \cdots w_k}) - v_{w_1 w_2 \cdots w_k} \times m(\theta^c_{w_1 w_2 \cdots w_k}) \geq d_{w_1 w_2 \cdots w_k}.$$

510. Exercise. Justify this observation and conclude that

$$E_{\tilde{\mu}}\left(\sum_{i=1}^{n} X_i(x) | x \in S'\right) \geq \sum_{k=0}^{n-1} \sum_{a_1 a_2 \cdots a_k \in i(S)} d_{a_1 a_2 \cdots a_k}$$
$$\times P(w_i = a_i, 1 \leq i \leq k, w_1 w_2 \cdots w_n \in S). \quad \bullet$$

We have not yet used the fact that $\overline{d}(\mu_n, \mu_n^{(S)}) > \epsilon$. We construct a coupling between μ_n and $\mu_n^{(S)}$ inductively, as in Example 301. Here's the basic idea: assume that we already know how the first k coordinates are coupled. Consider such a coupled pair of k-tuples $(\lambda_1, \lambda_2, \ldots, \lambda_k), (\gamma_1, \gamma_2, \ldots, \gamma_k)$. Conditioned on that coupled pair, we want to determine the joint distribution of $(\lambda_{k+1}, \gamma_{k+1})$. Note that the marginal distributions of λ_{k+1} and γ_{k+1} are μ and $\nu_{\gamma_1 \gamma_2 \cdots \gamma_k}$ respectively. Couple these two distributions as closely as possible. It follows that the probability that the coordinates don't match is $d_{\gamma_1 \gamma_2 \cdots \gamma_k}$.

511. Exercise. Do the coupling in this manner. Write down an expression for the expected mean Hamming distance between a pair of coupled words. $\quad \bullet$

Here are some more details concerning the foregoing exercise. Begin by letting P_1 be a coupling of μ with $\nu_{\emptyset}$ having the property that $E_{P_1}(d(\lambda, \gamma)) = d_{\emptyset}$. (Here d is the mean Hamming distance. Since it's being applied to a pair of one-letter words, its value is just 0 if the letters coincide and 1 if they don't.) Having constructed P_k (we are following the notation of Example 301; see in particular the footnotes there), for every fixed $\lambda_1, \ldots, \lambda_k \in \Lambda$ and $\gamma_1 \gamma_2 \cdots \gamma_k \in i(S)$, use μ for the first coordinate probability law and $\nu_{\gamma_1 \gamma_2 \cdots \gamma_k}$ for the second coordinate probability law conditioned on the occurrence of $\lambda_1, \ldots, \lambda_k$ and $\gamma_1, \ldots, \gamma_k$ at prior stages. (In the notation of the footnotes to Example 301, each $\mu_{\lambda_1, \ldots, \lambda_k}^{\gamma_1, \ldots, \gamma_k}$ is μ and each $\nu_{\lambda_1, \ldots, \lambda_k}^{\gamma_1, \ldots, \gamma_k}$ is $\nu_{\gamma_1 \gamma_2 \cdots \gamma_k}$.) Now when μ is coupled to $\nu_{\gamma_1 \gamma_2 \cdots \gamma_k}$, choose a coupling $P = P_{\gamma_1 \gamma_2 \cdots \gamma_k}$ having the property that $E_P(d(\lambda, \gamma)) = d_{\gamma_1 \gamma_2 \cdots \gamma_k}$. Finally let $P_{k+1}(\lambda_1, \ldots, \lambda_{k+1}, \gamma_1, \ldots, \gamma_{k+1}) = P_k(\lambda_1, \ldots, \lambda_k, \gamma_1, \ldots, \gamma_k) P_{\gamma_1 \gamma_2 \cdots \gamma_k}(\lambda_{k+1}, \gamma_{k+1})$. Continue in this fashion until P_n has been constructed.

512. Claim. For $1 \leq m \leq n$,

$$m E_{P_m}(d(b_1 b_2 \cdots b_m, a_1 a_2 \cdots a_m))$$
$$= \sum_{k=0}^{m-1} \sum_{a_1 a_2 \cdots a_k \in i(S)} d_{a_1 a_2 \cdots a_k} P(w_i = a_i, 1 \leq i \leq k, w_1 w_2 \cdots w_n \in S).$$

513. Exercise. Show that the truth of this claim for $m = n$, together with Exercise 510 and the fact that $\overline{d}(\mu_n, \mu_n^{(S)}) > \epsilon$, suffices for the proof of Theorem 501. $\quad \bullet$

We now proceed to the proof of Claim 512. To illustrate the idea, we'll justify the claim for $m = 2$ and leave the general case as an exercise. One has

$$E_{P_2}\big(\bar{d}(b_1 b_2, a_1 a_2)\big) = \sum_{b_1, b_2 \in \Lambda, a_1 a_2 \in i(S)} P_2(b_1, b_2, a_1, a_2) |\{i : b_i \neq a_i, 1 = i, 2\}|$$

$$= \sum_{b_1, b_2 \in \Lambda, a_1 a_2 \in i(S)} P_1(b_1, a_1) P_{b_1}(a_2, b_2) |\{i : b_i \neq a_i, 1 = i, 2\}|.$$

We can further analyze this expression to

$$\sum_{b_1, b_2 \in \Lambda, a_1 a_2 \in i(S), a_1 \neq b_1} P_1(b_1, a_1) P_{b_1}(a_2, b_2) + \sum_{b_1, b_2 \in \Lambda, a_1 a_2 \in i(S), a_2 \neq b_2} P_1(b_1, a_1) P_{b_1}(a_2, b_2)$$

$$= \sum_{b_1 \in \Lambda, a_1 \in i(S), a_1 \neq b_1} P_1(b_1, a_1) \left(\sum_{b_2 \in \Lambda, a_1 a_2 \in i(S)} P_{b_1}(a_2, b_2) \right)$$

$$+ \sum_{b_1 \in \Lambda, a_1 \in i(S)} P_1(b_1, a_1) \left(\sum_{b_2 \in \Lambda, a_1 a_2 \in i(S), a_2 \neq b_2} P_{b_1}(a_2, b_2) \right).$$

But this is just

$$\sum_{b_1 \in \Lambda, a_1 \in i(S), a_1 \neq b_1} P_1(b_1, a_1) + \sum_{a_1 \in i(S)} \left(\sum_{b_1 \in \Lambda} P_1(b_1, a_1) \right) d_{a_1}$$

$$= d_\emptyset + \sum_{a_1 \in i(S)} P(w_1 = a_1, w_1 w_2 \cdots w_n \in S) d_{a_1},$$

as required.

514. Exercise. Prove the general case of the claim, completing the proof of Theorem 501. $\square$

There are two things left to do to complete Step 3. First, we must show that independent concatenations are extremal. Then, we must show that the class of extremal processes is closed under dbar limits.

515. Theorem. *Independent concatenations are completely extremal.*

Idea of proof. This is a fairly straightforward application of Theorem 501. The basic idea is that an independent concatenation of period k on an alphabet Λ is very much like an independent process on the alphabet Λ^k. (So the letters are words of length k.)

516. Exercise. The reader should attempt to work out the details before, or instead of, reading the sketch given below. $\bullet$

Sketch of proof. Let $(X_i)_{i=-\infty}^\infty$ be a period k independent concatenation on a finite alphabet Λ induced by $(Y_i)_{i=-\infty}^\infty$. Let $\Lambda' = \Lambda^k$ and define a (stationary) process (U_i) on the finite alphabet Λ' by $U_i = X_{ik+1}X_{ik+2}\cdots X_{ik+k}$. For $n \in \mathbf{N}$, let μ_n be the measure induced on Λ^n by $(X_i)_{i=-\infty}^{\infty}$ and let γ_n be the measure induced on $\Lambda'^n = \Lambda^{kn}$ by $(U_i)_{i=-\infty}^{\infty}$. Let $\epsilon > 0$ and choose $\delta > 0$ and $N > \frac{1}{\epsilon}$ as in the definition of extremal for the independent, and hence completely extremal, process $(U_i)_{i=-\infty}^{\infty}$. Suppose now that $n \geq kN$ and suppose that ν is a submeasure of μ_n of size at least $2^{\frac{-\delta}{2k}n}$. Write $n = kn' + r$, where $0 \leq r < k$. ν induces a measure ν' on $\Lambda^{kn'} = \Lambda'^{n'}$ in the natural way ($nu'(w')$ is just the sum of $\nu(w)$, where w ranges over words of length n that begin with w'). It's easy to verify that $s(\frac{\nu'}{\gamma_{n'}}) \leq s(\frac{\nu}{\mu_n}) \leq 2^{\delta n} \leq 2^{\delta n'}$, so, since $n' \geq N, \overline{d}(\gamma_{n'}, \nu') \leq \epsilon$.

That is to say, there is a coupling of $\gamma_{n'}$ with ν' such that the average mean Hamming distance between coupled pairs is at most ϵ. Now, this coupling couples pairs of length n' words on the alphabet Λ', which can be identified with pairs of length kn' words on the alphabet Λ. The short description of this is that the coupling can be taken to be a coupling of $\mu_{kn'}$ and the measure $\nu_{kn'}$ that ν induces on length kn' words. Moreover, when the coupling is interpreted in this way, the average mean Hamming distance between coupled pairs cannot increase (though it certainly can decrease). All that remains is to inductively extend the coupling to a coupling of μ_n and ν_n using independent joinings of the conditionals on the remaining r times. Since $r < k < \epsilon kn'$, this can increase the average mean Hamming distance between coupled pairs by at most something like ϵ. Hence we have found a coupling of μ_n and ν such that the average mean Hamming distance between coupled pairs is at most 2ϵ, which is good enough.

517. Exercise. Fill in the details of this sketch.   □

518. Notation. Let X and Y be sets and consider a subset $C \subset X \times Y$. For $x \in X$, we write C_x to denote the set $C \cap \{x\} \times Y$.

519. Theorem. *The class of extremal processes is closed under dbar limits.*

Idea of proof. Suppose (S_i) is a sequence of extremal processes which converge in dbar to S. Select S_m close to S. Regard S_m and S to be measures on words of length n, where n is large. Let μ be a good dbar match, i.e. a measure on the product space with S and S_m as marginals. There are two small bad sets to consider.

There is a set Θ in the product space with μ small measure, such that off that set every ordered pair has coordinates that are close in the mean Hamming metric. And there is a set α of small S_m measure, such that every exponentially fat submeasure of S_m whose support is off α is close in dbar to S_m.

Let us refer to $\{(x, y) : y \notin \alpha$ and $(x, y) \notin \Theta\}$ as the *good* set.

There is an S-small set θ, such that for any $x \in \theta^c$, $\mu(\{(x, y) : y \in \Lambda^n, (x, y)$ is good$\} \setminus \{x\} \times \Lambda^n)$ is close to 1.

Fix an exponentially S-fat set F. (Here we restrict to subsets rather than submeasures, expecting the reader to generalize; see the proof sketch below for more details.) Let ν be the normalized probability measure on $F \times \Lambda^n$ induced by μ. ν is a coupling, in fact a good mean Hamming match, between F and its second coordinate projection ν_2. ν_2 is just as exponentially fat as F, and most of it is off α, so it can be changed slightly to get a measure ν_2' that is completely off α. F is close to ν_2 is close to ν_2' is close to S_m is close to S in dbar.

520. Abbreviation. We wish to show that approximating S with extremal S_m tends to make S extremal. After dodging obnoxious sets, we extract from the dbar coupling of S and S_m a dbar coupling of a prechosen fat submeasure of S with a fat submeasure of S_m.

521. Comment. The reader should note one can't use the above proof to show that a dbar limit of completely extremal processes is completely extremal.

Here are some more details.

Sketch of proof. Let $(S_i^{(j)})_{i=-\infty}^{\infty}$ be extremal processes on a finite alphabet Λ, $j \in \mathbf{N}$, and let $(S_i)_{i=-\infty}^{\infty}$ be a process on Λ such that $\lim_{j \to \infty} \overline{d}((S_i^{(j)})_{i=-\infty}^{\infty}, (S_i)_{i=-\infty}^{\infty}) = 0$. We must show that $(S_i)_{i=-\infty}^{\infty}$ is extremal. For $n \in \mathbf{N}$, let μ_n be the measure on Λ^n induced by $(S_i)_{i=-\infty}^{\infty}$ and let $\mu_n^{(j)}$ be the measure on Λ^n induced by $(S_i^{(j)})_{i=-\infty}^{\infty}$.

Let $\epsilon > 0$. We shall find $\delta > 0$ and N such that for every $n < N$ there is a set $B_n \subset \Lambda^n$ with $\mu_n(B_n) < \epsilon$ and such that for any measure ν on Λ^n with $\nu(B_n) = 0$ and $s(\frac{\nu}{\mu_n}) \leq 2^{\delta n}$, one has $\overline{d}(\mu_n, \nu) < 3\epsilon$.

Choose j with $\overline{d}((S_i^{(j)})_{i=-\infty}^{\infty}, (S_i)_{i=-\infty}^{\infty}) < \frac{\epsilon^3}{10}$. Since $(S_i^{(j)})_{i=-\infty}^{\infty}$ is extremal, we can choose $\delta > 0$ and N having the property that for every $n > N$ there is a set $B_n^{(j)} \subset \Lambda^n$ with $\mu_n^{(j)}(B_n^{(j)}) < \frac{\epsilon^2}{2}$ and such that for any submeasure ν of $\mu_n^{(j)}$ of size at least $2^{-\delta n}$ satisfying $\nu(B_n^{(j)}) = 0$, $\overline{d}(\mu_n^{(j)}, \nu) < \frac{\epsilon}{10}$.

Let $n > N$ and pick a coupling μ of μ_n and $\mu_n^{(j)}$, that is a measure on $\Lambda^n \times \Lambda^n$, having the property that the average mean Hamming distance between coupled pairs of words is less than $\frac{\epsilon^3}{8}$. Let Θ be the set of pairs $(w_1, w_2) \in \Lambda^n \times \Lambda^n$ having the property that $d(w_1, w_2) > \frac{\epsilon}{4}$. Then $\mu(\Theta) \leq \frac{\epsilon^2}{2}$. We denote by G the set of pairs (w_1, w_2) such that $(w_1, w_2) \notin \Theta$ and $w_2 \notin B_n^{(j)}$. Clearly $\mu(G) > 1 - \epsilon^2$.

522. Exercise. Show that the set $B_n = \{w \in \Lambda^n : \frac{\mu(G_w)}{\mu_n(w)} \leq 1 - \epsilon\}$ satisfies $\mu_n(B_n) \leq \epsilon$. *Hint: note that $\int \frac{\mu(G_w^c)}{\mu_n(w)} d\mu_n(w) = \mu(G^c) \leq \epsilon^2$.* •

Now let ν be a measure on Λ^n such that $\nu(B_n) = 0$ and $s(\frac{\nu}{\mu_n}) \leq 2^{\delta n}$. We will show that $\overline{d}(\mu_n, \nu) < 3\epsilon$, thus completing the proof.

Define a measure μ' on $\Lambda_n \times \Lambda_n$ by the rule $\mu'(w_1, w_2) = 0$ if $\mu_n(w_1) = 0$ and $\mu'(w_1, w_2) = \mu(w_1, w_2) \cdot \frac{\nu(w_1)}{\mu_n(w_1)}$ otherwise. It is easy to show that μ' is a probability measure and that the projection of μ' onto the first coordinate is ν. Notice also that $s(\frac{\mu'}{\mu}) \leq 2^{\delta n}$. Let $\nu^{(j)}$ be the projection of μ' onto the second coordinate. That is, $\nu^{(j)}(w_2) = \sum_{w_1} \mu'(w_1, w_2)$. μ' is a coupling of ν and $\nu^{(j)}$. One sees that $\mu'(\Theta) \leq e$, from which it immediately follows that $\overline{d}(\nu, \nu^{(j)}) \leq \frac{5\epsilon}{4}$. Moreover, one has $s(\frac{\nu^{(j)}}{\mu_n^{(j)}}) \leq 2^{\delta n}$.

523. Exercise. Show that $\nu^{(j)}(B_n^{(j)}) \leq \epsilon$, and that consequently there exists a measure ν' on Λ^n differing from $\nu^{(j)}$ by at most ϵ in the variation distance such that $\nu'(B^{(j)}) = 0$ and such that $s(\frac{\nu'}{\mu_n^{(j)}}) \leq 2^{\delta n}$. Conclude that $\overline{d}(\mu_n^{(j)}, \nu') < \frac{\epsilon}{10}$. •

This finishes the proof, as by the triangle inequality,

$$\overline{d}(\mu_n, \nu) \leq \overline{d}\left(\mu_n, \mu_n^{(j)}\right) + \overline{d}\left(\mu_n^{(j)}, \nu'\right) + \overline{d}\left(\nu', \nu^{(j)}\right) + \overline{d}\left(\nu^{(j)}, \nu\right)$$

$$\leq \frac{\epsilon^3}{8} + \frac{\epsilon}{10} + \epsilon + \frac{5\epsilon}{4} < 3\epsilon.   \qquad \square$$

5.4. Step 4: FD ⊂ IC

This step requires a bit of work but is fairly straightforward. We start with a definition.

524. Definition. Let $(X_i)_{i=-\infty}^{\infty}$ be a stationary process and let $\epsilon > 0$. Let $(Y_i)_{i=-\infty}^{\infty}$ be the independent stationary process on $\{0, 1\}$ with $P(Y_i = 1) = \epsilon$. We define a new process $(Z_i)_{i=-\infty}^{\infty}$ as follows. Select i such that $Y_i = 1$. Let j be the least number greater than i such that $Y_j = 1$. For each such i, j we let $Z_{i+1}, Z_{i+2}, \ldots, Z_j$ have the same joint distribution as $X_1, X_2, \ldots X_{j-1}$, but require that $Z_{i+1}, Z_{i+2}, \ldots, Z_{j-1}$ be independent of the other Z_is. $(Z_i)_{i=-\infty}^{\infty}$ is called the ϵ-*startover process* of $(X_i)_{i=-\infty}^{\infty}$.[84]

[84] Some more details. We can assume that $(X_i)_{i=-\infty}^{\infty}$ and $(Y_i)_{i=-\infty}^{\infty}$ are defined on the same space Z and are independent. To carry out the construction, for each pair of integers $i < j$ define families of random variables $S^{(i,j)} = \left\{R_k^{(i,j)} : i < k \leq j\right\}$, in such a way that $(R_{i+1}, R_{i+2}, \ldots, R_j)$ is distributed as $(X_1, X_2, \ldots, X_{j-i})$ is, and such that, moreover, the families $S^{(i,j)}$ are mutually independent and independent of $(Y_i)_{i=-\infty}^{\infty}$. For a.e. $z \in Z$, $Y_i(z) = 1$ for both arbitrarily large positive and negative i. For such z, if $Y_i(z) = 1 = Y_j(z)$ and $Y_k \neq 1, i < k < j$, one puts $Z_k(z) = R_k^{(i,j)}(z), i < k \leq j$.

525. Exercise. ϵ-startover processes are stationary.

526. Comment. Intuitively, at each time step the ϵ-startover process runs like the X process with probability $1 - \epsilon$, starting over with probability ϵ.

527. Theorem. ϵ-*startover processes are* **IC**.

Idea of proof. Let X be a process and Z its ϵ-startover process. Let m be large enough so that with very high probability Z has started over by time m. Let $n \gg m$. We independently concatenate the distribution of $X_1, X_2, X_3, \ldots, X_n$ to form $Y = (Y_1, Y_2, Y_3, \ldots, Y_n)(Y_{n+1}, Y_{n+2}, Y_{n+3}, \ldots, Y_{2n})(Y_{2n+1}, Y_{2n+2}, Y_{2n+3}, \ldots, Y_{3n}) \ldots$ Now we couple Y to Z as follows.

 (i) First couple times 1 to $n - m$, $n + 1$ to $2n - m$, $2n + 1$ to $3n - m$, etc.
 (ii) Conditioned on how we coupled the above times we couple the remaining times independently.

To accomplish (i) couple each set of times $\{kn + 1, \ldots, (k + 1)n - m\}$ independently if X did not start over during times $\{kn - m, \ldots, kn\}$ and identically the same if X did start over during $\{kn - m, \ldots, kn\}$.

Sketch of proof. Let $\epsilon > 0$, let $(X_i)_{i=-\infty}^{\infty}$ be a stationary process on a finite alphabet Λ and let $(Z_i)_{i=-\infty}^{\infty}$ be its ϵ-startover process. Let $(Y_i)_{i=-\infty}^{\infty}$ be as in the construction of the ϵ-startover process. Let $\delta > 0$ be arbitrary and choose a positive integer m so large that $(1 - \epsilon^2)^m < \delta$. Choose $n > \frac{m}{\delta}$. Let $(U_i, W_i)_{i=-\infty}^{\infty}$ be an independent concatenation[85] of period n induced by $(Z_i, Y_i)_{i=-\infty}^{\infty}$. In particular, $(U_1, \ldots, U_n, W_1, \ldots, W_n)$ is to have the same distribution as $(Z_1, \ldots, Z_n, Y_1, \ldots, Y_n)$. Notice that $(Y_i)_{i=-\infty}^{\infty}$ encodes the startover times of $(Z_i)_{i=-\infty}^{\infty}$ and $(W_i)_{i=-\infty}^{\infty}$ encodes the startover times of $(U_i)_{i=-\infty}^{\infty}$.[86]

528. Exercise. It suffices for the proof to show that $(U_i)_{i=-\infty}^{\infty}$ is within 4δ of $(Z_i)_{i=-\infty}^{\infty}$ in dbar. $\qquad\bullet$

As the previous exercise suggests, the thing to do now is to couple $(Z_i)_{i=-\infty}^{\infty}$ to $(U_i)_{i=-\infty}^{\infty}$. In Theorem 485, the idea was to isolate blocks encompassing a large proportion of the times on which the two processes had identical marginals, so the diagonal coupling could be used in the induction step. Here, things are a bit more complicated. As in the prior proof, we couple blocks of times $\{kn + 1, kn + 2, \ldots, kn + n\}$ in turn for $k = 0, 1, 2, 3, \ldots$ Inside the blocks, however, we do something a bit different, which actually

---

[85] The process pairs we are considering here can be viewed as processes on the alphabet $\Lambda \times \{0, 1\}$.
[86] Of course, U_i has additional startover times occurring with period n.

doesn't even fall precisely under "coupling by induction" as we introduced it in Example 301.

Now to the details: let us say $(Z_i, Y_i)_{i=-\infty}^{\infty}$ has been coupled to $(U_i, W_i)_{i=-\infty}^{\infty}$ over several n-blocks of times, and we want now to extend the coupling over the block $\{kn + 1, kn + 2, \ldots, kn + n\}$. First, we count our way through the block using the independent coupling *on the second coordinate only.*[87] That is, we couple just (Y_i) and (W_i) on this first run-through. Naturally, everything is independent of previous blocks here as well.

529. Exercise. Make sure you understand what has happened so far.   •

The next stage is to run through the block again, coupling the first coordinate. Note that the next-letter probabilities are conditioned on everything previously coupled, which includes not only previous blocks, but also the startover times over the current block, as encoded by (Y_i) and (W_i). Let us say that these startover times are given in a specific instance by words (y_i) and (w_i), $kn + 1 \le i \le kn + n$. Let i_{min} be the least i in the block (if any) such that $y_i = w_i = 1$. Now, when we are coupling the first coordinate through the current block, conditioned on everything previously coupled, we use the independent joining at times $kn + 1, kn + 2, \ldots, i_{min}$ and the diagonal joining thereafter.

530. Exercise. Justify the argument given to this stage, namely that the diagonal joining can be used between simultaneous startovers within the block. Say why it cannot be used elsewhere within the block. *Hint: the marginals may depend on previously coupled blocks.*

One now completes the proof by noting that, with probability at least $1 - \delta$, there is a simultaneous startover time in the first m-sub-block of the n-block we are considering, which allows us to use the diagonal coupling on all but at most m of the n coordinates of the block, where $\frac{m}{n} < \delta$.

531. Exercise. Give the details.   □

532. Theorem. *Let $(X_i)_{i=-\infty}^{\infty}$ be a stationary process on a finite alphabet and let H be its entropy. If $\delta > 0$, then for all small enough $\epsilon > 0$, if $Z = (Z_i)_{i=-\infty}^{\infty}$ is the ϵ-startover process for $(X_i)_{i=-\infty}^{\infty}$, one has $|H(Z) - H| < \delta$.*

Idea of proof. Fix a big n such that the number of reasonable n-names is about 2^{nH}. Now pick $\epsilon \ll \frac{1}{n}$, so that the chances of $(Z_i)_{i=-\infty}^{\infty}$ starting over

---

[87] The reader may want to check that we could use the diagonal coupling here. If we did that, then later we wouldn't have to worry about simultaneous startovers, since they would all be simultaneous. It doesn't really matter.

in a typical n-block are miniscule. Now for large k, if you look at a kn-word in $(Z_i)_{i=-\infty}^{\infty}$ and break it into k n-words, most of those n-words will be reasonable for X most of the time. The rest is just counting.   □

533. Theorem. FD ⊂ IC.

Idea of proof. Just observe that, restricted to a large enough initial block of times, the ϵ-startover process of (X_i) is close to (X_i) in distribution and entropy. Apply Theorem 527.

Sketch of proof. Let $X = (X_i)_{i=-\infty}^{\infty}$ be a finitely determined process on a finite alphabet Λ and let $\alpha > 0$ be arbitrary. Choose, by the definition of **FD**, $n \in \mathbf{N}$ and $\delta > 0$ such that for any ergodic process $Y = (Y_i)_{i=-\infty}^{\infty}$ on Λ, if $v(\mu_{X,n}, \mu_{Y,n}) < \delta$ and $|H(X) - H(Y)| < \delta$ then $\overline{d}(X, Y) < \alpha$. Now choose $\epsilon > 0$ small enough that the ϵ-startover process $Z = (Z_i)_{i=-\infty}^{\infty}$ for $(X_i)_{i=-\infty}^{\infty}$ satisfies $|H(X) - H(Z)| < \delta$ and $v(\mu_{X,n}, \mu_{Z,n}) < \delta$. (The first requirement may be satisfied using Theorem 532, the second is a simple exercise.) One may conclude that $\overline{d}(X, Z) < \alpha$. Since α is arbitrary and Z is **IC** by Theorem 527, we are done.   □

5.5. Step 5: EX ⊂ VWB

Let $(Y_i)_{i=-\infty}^{\infty}$ be a stationary process on a finite alphabet Λ. As usual, we assume the space of the random variables to be $\Omega = \Lambda^{\mathbf{Z}}$, and denote by μ the measure induced by $(Y_i)_{i=-\infty}^{\infty}$ on Ω. Recall that we denote the σ-algebra generated by $(Y_i)_{i=-\infty}^{-1}$ by *past*; we extend this notion. For each $n \in \mathbf{Z}$, denote the σ-algebra generated by $(Y_i)_{i=-\infty}^{n}$ by $\mathcal{A}_n$. (So that *past* $= \mathcal{A}_{-1}$.) Next, denote by μ_n the restriction of μ to $\mathcal{A}_n$. Finally let $\Omega_n = \Lambda^{\{\ldots, n-2, n-1, n\}}$. In a small abuse of notation, we may speak of the probability spaces $(\Omega_n, \mathcal{A}_n, \mu_n)$ rather than $(\Omega, \mathcal{A}_n, \mu_n)$; we hope this doesn't alarm the reader too greatly.

534. Exercise. Note that $\mathcal{A}_n \subset \mathcal{A}_{n+1}$. Show that the disintegration of μ_{n+1} over μ_n is given by $\int_{\Omega_{n+1}} d\mu_{n+1} = \int_{\Omega_n} \int_{\Lambda} d\mu_y \, d\mu_n(y)$, where, for $y = (y_i)_{i=-\infty}^{n}$, $\mu_y(\lambda) = P(Y_{n+1} = \lambda | X_i = y_i, i \leq n)$.

535. Let $m \in \mathbf{N}$. For $a_0 \cdots a_{m-1} \in \Lambda^m$ and $y = (y_i)_{i=-\infty}^{\infty} \in \Omega$ write $\gamma_y(a_0 \cdots a_{m-1}) = P(Y_i = a_i, 0 \leq i < m | past)(y) = P(Y_i = a_i, 0 \leq i < m | Y_i = y_i, i < 0)$. γ_y is a measure on the finite space Λ^m and we denote its partition entropy (see Comment 396) by $H_m(y)$.

536. Definition. The function $y \to H_m(y)$ is called the *conditioned entropy of the m future given the past*.

537. Theorem. $E(H_m) = mH$, *where* H *is the entropy of the process* $(Y_i)_{i=-\infty}^{\infty}$.

Idea of proof. The easiest proof is to show that $E(H_m(y))$ is the limit as $n \to \infty$ of $H(X_i, 1 \le i \le m | X_i, -n \le i \le 0)$. Apply Theorem 425. In the sketch we offer a different proof.

Sketch of proof. The proof we just suggested, while being rather simple, may not give all readers much sense of what's going on. Here we'll outline a more direct proof. As the notation is somewhat involved, we limit ourselves to the case $m = 2$, leaving the general case to the reader. The first step is to show the following:

Claim: $E\big(- \log P(Y_0 Y_1 | past) \big) = 2E \big(- \log P(Y_0 | past) \big) = 2H.$

The second equality is just Corollary 379. We explain the notation: $P(Y_0 Y_1 | past)$ is a Ω_1-measurable random variable whose value at a.e. $y = (y_i)_{i=-\infty}^{\infty}$ is given by

$$P(Y_0 = y_0, Y_1 = y_1 | Y_i = y_i, i < 0) = P(Y_0 = y_0 | Y_i = y_i, i < 0)$$
$$\times P(Y_1 = y_1 | Y_i = y_i, i < 1).$$

It follows that

$$- \log P(Y_0 Y_1 | past)(y) = - \log P(Y_0 = y_0 | Y_i = y_i, i < 0)$$
$$- \log P(Y_1 = y_1 | Y_i = y_i, i < 1)$$
$$= - \log P(Y_0 | \mathcal{A}_{-1}) - \log P(Y_1 | \mathcal{A}_0).$$

(Recall that $past = \mathcal{A}_{-1}$.) Taking expectations, the claim therefore follows from:

538. Exercise. For any $n \in \mathbf{Z}$, one has $E\big(- \log P(Y_{n+1} | \mathcal{A}_n) \big) = E\big(- \log P(Y_0 | past) \big).$ •

Now we start disintegrating the measure μ_1. Notation here is as follows. $y = (y_i)_{i=-\infty}^{-1} \in \Omega_{-1}$, $y_0, y_1 \in \Lambda$, so that $yy_0 \in \Omega_0$ and $yy_0y_1 \in \Omega_1$. One has

$$2H = \int_{\Omega_1} - \log P(Y_0 Y_1 | past)(yy_0 y_1) \, d\mu_1(yy_0 y_1)$$
$$= \int_{\Omega_0} \int_\Lambda - \log P(Y_0 Y_1 | past)(yy_0 y_1) \, d\mu_{yy_0}(y_1) \, d\mu_0(yy_0)$$
$$= \int_{\Omega_{-1}} \int_\Lambda \int_\Lambda - \log P(Y_0 Y_1 | past)(yy_0 y_1) \, d\mu_{yy_0}(y_1) \, d\mu_y(y_0) \, d\mu_{-1}(y).$$

We can rewrite the interior double integral as

$$\sum_{y_0, y_1 \in \Lambda} \left(-\log P(Y_0 Y_1 | past)(yy_0 y_1) \right) \mu_{yy_0}(y_1) \mu_y(y_0).$$

But

$$\mu_{yy_0}(y_1) \mu_y(y_0) = P(Y_1 | A_0)(yy_0 y_1) P(Y_0 | A_{-1})(yy_0) = P(Y_0 Y_1 | A_{-1})(yy_0 y_1)$$
$$= \gamma_y(y_0 y_1),$$

so the double integral is just the partition entropy of γ_y, namely $H_2(y)$.

539. Exercise. Extend the above argument to general m. $\square$

The next couple of exercises are technical in nature. They or something like them are needed to show that for a stationary process, most pasts induce conditioned futures that are exponentially fat submeasures of the unconditioned future.

540. Exercise. Let Λ be a finite alphabet, $m \in \mathbf{N}$ and $\epsilon > 0$, let ν be a measure on Λ^m and let $R \subset \Lambda^m$ with $\nu(R) \le \epsilon$. Then the contribution of the names in R to the partition entropy of ν is at most $-\epsilon \log \epsilon + \epsilon m \log |\Lambda|$. Indeed, there is some measure ν' that is supported on R^c and is ϵ-close to ν in the variation metric such that $H(\nu') \ge H(\nu) - (-\epsilon \log \epsilon + \epsilon m \log |\Lambda|)$.

541. Important comment. We'd like to avoid messy expressions when we can. When ϵ is small enough (given $|\Lambda|$), $-\epsilon \log \epsilon + \epsilon m \log |\Lambda| \le m \epsilon^{\frac{2}{3}}$ (for example) gives a cleaner, if less precise, estimate.

542. Exercise. Suppose you have a measure μ on a finite space X. Let $m \in \mathbf{N}$, $\epsilon > 0$, $H > 0$. Suppose that X has enough atoms of μ-measure enough greater than $2^{-mH+\epsilon m}$ and enough other atoms of μ-measure less than 2^{-mH} that it is possible to move a total of ϵ of the measure from the first set of atoms to the second set of atoms without lowering the measure of any atom in the first set below $2^{-mH+\epsilon m}$ and without raising the measure of any atom in the second set above 2^{-mH}. Call the measure you get after the move ν. Show that $H(\nu) \ge H(\mu) + \frac{\epsilon^2 m}{\ln 2}$. *Hint: consider that $H(\mu) = \sum_x f(x)$ where $f(x) = -x \log x$. Compute the derivative f' and estimate $H(\nu) - H(\mu)$ with the mean value theorem.*

543. Theorem. *Let $(X_i)_{i=-\infty}^{\infty}$ be an ergodic stationary process on a finite alphabet Λ. For any $\epsilon > 0$, there exists M such that for all $m \ge M$ there is a set $S \subset \Omega_{-1}$ with $\mu_{-1}(S) \ge 1 - \epsilon$ such that for every $y \in S$, there is a measure γ on Λ^m such that*[88] $v(\gamma, \gamma_y) \le \epsilon$ *and such that $s\left(\frac{\gamma}{\mu_m}\right) \le 2^{\epsilon m}$.*

---

[88] Recall that γ_y is the probability law on the m future given that the past is y.

544. Abbreviation. If you select ϵ and let m be large enough, then for most (probability at least $1 - \epsilon$) pasts, the m future given the past is close (distance at most ϵ) in variation to an exponentially fat (size at least $2^{\epsilon m}$) submeasure of the unconditioned m future.

545. Further abbreviation. Most conditioned futures are nearly exponentially fat submeasures of the unconditioned future.

Idea of proof. Let θ be the set of all names that eventually become and stay reasonable. θ has measure 1 by the Shannon–McMillan–Breiman theorem, hence for all pasts y off a set of measure 0, the conditioned future given y assigns measure 1 to θ. Conditioned on such a past y, for sufficiently large m most (in the conditioned-on-y sense) m-names are reasonable (in the unconditioned sense).

Now it is an elementary fact that if you have a random variable with a finite expectation which nearly always takes on values at most slightly greater than its expectation and never takes on values much greater than its expectation, then it rarely takes on values much lower than its expectation. We apply this fact to $H_m(y)$, the conditioned entropy of the m future given the past. In particular:

(i) $E(H_m) = mH$,
(ii) H_m is rarely much bigger than mH because
    (a) usually most of the m future is on reasonable names so
    (b) usually by making only a small change in the m future you can get a measure which lives entirely on reasonable names.
    (c) Making this small change only slightly affects entropy.
    (d) The most entropy you can get if you live on reasonable names is when you assign every reasonable name the same measure.
    (e) If you do that you get about mH entropy because by Shannon–McMillan–Breiman there are about $2mH$ reasonable names.
(iii) $m \log |\Lambda|$ is an absolute upper bound for H_m.

These facts imply that the conditioned entropy of the m future given the past is rarely much smaller than mH. Use the conditioning property to show that when a measure lies almost entirely on reasonable names and has entropy not much smaller than mH then it is rare for a name to have measure much bigger than 2^{-mH}. Alter such a measure to live only on reasonable names whose measure is not much bigger than 2^{-mH}. If you then multiply their measures by $2^{-\epsilon m}$ you will get only reasonable names with measures not more than reasonable.

Sketch of proof. Let $1 \gg \epsilon > 0$ and let H be the entropy of $(X_i)_{i=-\infty}^{\infty}$. Put $\delta = \frac{\epsilon^{20}}{10^6}$. Let μ be the measure on $\Omega = \Lambda^{\mathbf{Z}}$ induced by $(X_i)_{i=-\infty}^{\infty}$. Let

$$\theta = \left\{ x = (x_i)_{i=-\infty}^{\infty} : \lim_{n \to \infty} -\frac{1}{n} \log P(X_i - x_i, 1 \le i \le n) = H \right\}.$$

For $n \in \mathbf{N}$, put

$$\theta_n = \left\{ x = (x_i)_{i=-\infty}^{\infty} : \text{for all } m \ge n, \left| -\frac{1}{m} \log P(X_i - x_i, 1 \le i \le m) - H \right| \le \delta \right\}.$$

Then $\theta \subset \bigcup_{n=1}^{\infty} \theta_n$. One has $\mu(\theta) = 1$, which implies that $\int_{\Omega_{-1}} \mu_y(\theta) \, d\mu_{-1}(y) = 1$. Hence letting $G = \{ y \in \Omega_{-1} : \mu_y(\theta) = 1 \}$, one has $\mu_{-1}(G) = 1$. For $y \in G$, choose $m(y)$ so large that $\mu_y(\theta_{m(y)}) \ge 1 - \delta$.

546. Exercise. Let $R_n = \{ (x_i)_{i=1}^{n} : (x_i)_{i=-\infty}^{\infty} \in \theta_n \}$. Show that $|R_n| \le 2^{mH + m\delta}$. $\qquad \bullet$

547. Exercise. If $m \ge m(y)$ then $H_m(y) \le mH + m\delta^{\frac{2}{3}}$. *Hint: use the fact that* $\gamma_y(R_n) \ge 1 - \epsilon$ *and see Comment 541 above.* $\qquad \bullet$

Choose m such that $\mu_{-1}(\{ y \in \Omega_{-1} : m(y) \le m \}) \ge 1 - \delta$. Let now $E_1 = \{ y \in \Omega_{-1} : H_m(y) \ge mH \}$ and let $E_2 = \{ y \in \Omega_{-1} : H_m(y) < mH \}$. Since $E(H_m) = mH$, one has $\int_{E_1} H_m(y) - mH \, d\mu_{-1}(y) = \int_{E_2} mH - H_m(y) \, d\mu_{-1}(y)$.

548. Exercise. Show that $\int_{E_1} H_m(y) - mH \, d\mu_{-1}(y) \le m\delta^{\frac{2}{3}} + m\delta \log |\Lambda| \le m\delta^{\frac{1}{2}}$. $\qquad \bullet$

From the preceding exercise, we get that $\int_{E_2} mH - H_m(y) \, d\mu_{-1}(y) \le m\delta^{\frac{1}{2}}$, so that the set $B = \left\{ y : mH - H_m(y) \ge m\delta^{\frac{1}{4}} \right\}$ satisfies $\mu_{-1}(B) \le \delta^{\frac{1}{4}}$. Let $S = \left\{ y : H_m(y) \ge mH - m\delta^{\frac{1}{4}} \text{ and } m \ge m(y) \right\}$. Note that for any $y \in S$, $\mu_y(\theta_m) \ge 1 - \delta$. This implies that $\gamma_y(R_m) \ge 1 - \delta$, and of course $H_m(y) \ge mH - m\delta^{\frac{1}{4}}$. By Exercise 541, less than $m\delta^{\frac{2}{3}}$ of the entropy of γ_y comes from R_m^c. Therefore, we may find a measure γ_y' supported on R_m with $\upsilon(\gamma_y, \gamma_y') \le \delta$, such that the entropy of γ_y' is at least $mH - m\delta^{\frac{1}{8}}$.

Suppose now, for a contradiction, that for no measure γ supported on R_m with $\upsilon(\gamma, \gamma_y') \le \frac{\epsilon}{2}$ does one have $s(\frac{\gamma}{\mu_m}) \le 2^{\epsilon m}$. In particular, one cannot have for such γ that $\gamma(w) \le 2^{-mH + \delta m + \frac{\epsilon}{2} m}$ for all $w \in R_m$. This means that one can move $\frac{\epsilon}{2}$ of measure off words having γ_y'-measure more than $2^{-mH + \delta m + \frac{\epsilon}{2} m}$ (without reducing any of these words to a measure below this level) to words in R_m of γ_y'-measure less than $2^{-mH + \delta m}$ (without raising any of these words to a measure above this level) to form a new measure γ^*. Denote the partition entropy of γ^* by H^*. By Exercise 542,

$H^* \geq H(\gamma_y') + \frac{\epsilon^2}{4}m \geq mH - m\delta^{\frac{1}{8}} + \frac{\epsilon^2}{4}m$. On the other hand, γ^* is supported on $\bar{R}_m$, so $H^* \leq mH + m\delta$. This is a contradiction.

549. Exercise. Shore up the details, completing the proof.   □

550. Exercise. Show that the proof of Theorem 543 can be modified to allow that if $C \subset \Lambda^m$ is small in the unconditioned m-future, then the measures γ_y can be chosen disjoint from C. To be precise, show: Let $(X_i)_{i=-\infty}^{\infty}$ be an ergodic stationary process on a finite alphabet Λ. For any $\epsilon > 0$, there exists M such that for all $m \geq M$ and all sets $C \subset \Lambda^m$ for which $\mu_m(C) \leq \frac{\epsilon^{20}}{10^6}$ there is a set $S \subset \Omega_{-1}$ with $\mu_{-1}(S) \geq 1 - \epsilon$ such that for every $y \in S$, there is a measure γ on Λ^m such that $\gamma(C) = 0$, $v(\gamma, \gamma_y) \leq \epsilon$ and such that $s(\frac{\gamma}{\mu_m}) \leq 2^{\epsilon m}$. (Abbreviation: most conditioned futures are nearly exponentially fat submeasures of the unconditioned future disjoint from a prechosen small set.) *Hint: in the proof, replace S by $S \cap \{y : \mu_y(C) < \delta\}$ and R_m by $R_m \cap C$.*

551. Theorem. *Any extremal process is ergodic.*

Idea of proof. We outline a proof of the contrapositive. Non-ergodicity implies the existence of a finite word whose frequency of occurrence in a randomly chosen $(x_i)_{i=-\infty}^{\infty} \in \Lambda^{\mathbb{Z}}$ is non-constant. Select $a < b$ such that both the event that the frequency of the word is less than a and the event that the frequency of the word is greater than b have positive probability.

552. Exercise. Show that these two events induce fat submeasures of μ_n, for large n, that cannot possibly be close in dbar.   □

553. Theorem. $EX \subset VWB$.

Sketch of proof. Let $(X_i)_{i=-\infty}^{\infty}$ be extremal. By Theorem 551, $(X_i)_{i=-\infty}^{\infty}$ is ergodic. To show that $(X_i)_{i=-\infty}^{\infty}$ is very weak Bernoulli, it is sufficient to exhibit, for arbitrary $\epsilon > 0$, some M such that for all $m \geq M$ one has $\mu_{-1}(\{y : \bar{d}(\gamma_y, \mu_m) > 2\epsilon\}) < \epsilon$. Let $\epsilon > 0$. Since $(X_i)_{i=-\infty}^{\infty}$ is extremal, there is some $\delta > 0$ (we'll assume $\delta < \epsilon$) and $N \in \mathbb{N}$ such that for all $n \geq N$, there is a set $B_n \subset \Lambda^n$ with $\mu_n(B_n) < \frac{\epsilon^{20}}{10^6}$ having the property that for any submeasure v of μ_n with $s(\frac{v}{\mu_n}) \leq 2^{\delta n}$ and $v(B_n) = 0$, $\bar{d}(\mu_n, v) \leq \epsilon$.

Since $(X_i)_{i=-\infty}^{\infty}$ is ergodic, by Exercise 550 there is an $M > N$ such that for every $m \geq M$ there is a set $S \subset \Omega_{-1}$ with $\mu_{-1}(S) \geq 1 - \delta$ and such that for all $y \in S$, there is some measure γ on Λ^m with $\gamma(B_m) = 0$, $v(\gamma, \gamma_y) \leq \delta$ and such that $s(\frac{\gamma}{\mu_m}) \leq 2^{\delta m}$. This latter implies by the above that $\bar{d}(\gamma, \mu_m) \leq \epsilon$ and hence that $\bar{d}(\gamma_y, \mu_m) \leq 2\epsilon$.   □

5.6. Step 6: EX ⊂ FD

Theorem 543 (also Exercise 550) says that sufficiently far out conditioned futures are nearly exponentially fat subsets of the unconditioned future, but it doesn't tell you how far out. We'll say a bit about this because we need a stronger version for this step. By examining the proof, one can see that as soon as most names are reasonable, that's sufficiently far. We now put the precise formulation we'll need in an exercise; the reader should be able to do the exercise by mimicking the proof of Theorem 543. There is one subtlety, however. In Theorem 543, one conditions on the whole past and uses the fact that $E(H_m) = mH$. In the exercise below, one needs an analogous fact that applies to the case where one conditions on a finite past. *Hint: use Theorem 421 to show that the expected entropy of the m-future given the n-past is non-increasing in n, so that if it is close to mH for n = 0, it must be close to mH for all n.*

554. Exercise. Let $(X_i)_{i=-\infty}^{\infty}$ be an ergodic stationary process on a finite alphabet Λ and having entropy H. Let $\epsilon > 0$ and $m \in \mathbf{N}$. Let μ_m be the measure on Λ^m induced by the process and let μ'_m be any measure with $v(\mu_m, \mu'_m) < \epsilon$. Put

$$R_m = \left\{ a = a_1 a_2 \cdots a_m \in \Lambda^m : \left| -\frac{1}{m} \log P(a_1 a_2 \cdots a_m) - H \right| < \epsilon \right\}.$$

Show that if $\mu_m(R_m) > 1 - \epsilon$ then for any $n \in \mathbf{N}$ and all sets $C \subset \Lambda^m$ with $\mu_m(C) \leq \epsilon$, there is a set $S \subset \Lambda^{\{-n+1, -n+2, \ldots, 0\}}$ with $P(X_{-n+1} X_{-n+2} \cdots X_0 \in S) \geq 1 - \epsilon^{\frac{1}{20}}$ such that for all $a \in S$, there is a measure γ on Λ^m such that $\gamma(C) = 0$, $v(\gamma, \gamma_a) \leq \epsilon^{\frac{1}{20}}$, where

$$\gamma_a(E) = P(X_1 X_2 \cdots X_m \in E | X_i = a_i, -n < i \leq 0),$$

and such that $s(\frac{\gamma}{\mu_m}) \leq 2^{m\epsilon^{\frac{1}{20}}}$.

555. Theorem. *EX ⊂ FD.*

Idea of proof. Let (X_i) be an extremal process. Pick m so that most m-names are reasonable, and also big enough so that off a small error set, all exponentially fat submeasures of μ_m are dbar-close to μ_m. Then pick a process $(Y_i)_{i=-\infty}^{\infty}$ that approximates $(X_i)_{i=-\infty}^{\infty}$ in both m-distribution and entropy. Notice that this implies in particular that most m-names for (Y_i) are reasonable. Now for either process, Exercise 554 says that conditioned on the n-past for any n, conditioned m-futures are usually close in variation to exponentially fat submeasures of the unconditioned m-future of the (X_i) process. It follows that they can each be coupled closely in dbar to that unconditional and hence to each other. Since this holds for any natural number n, you can just couple (X_i) closely to (Y_i) in dbar inductively, one m-block at a time. $\qquad\square$

5.7. Step 7: VWB $\subset$ IC

556. Theorem. *VWB $\subset$ IC.*

Idea of proof. This is easy. Start with a very weak Bernoulli process $(X_i)_{i=-\infty}^{\infty}$. We must couple it closely in dbar to an independent concatenation. Fix a large n such that the unconditioned n-future is close in dbar to the n-future conditioned on the past, for almost all pasts. If m is large enough, this is still true even if you only condition on the m-past. Let $(Z_i)_{i=-\infty}^{\infty}$ be the period-n independent concatenation induced by $(X_i)_{i=-\infty}^{\infty}$. Start coupling $(X_i)_{i=-\infty}^{\infty}$ and $(Z_i)_{i=-\infty}^{\infty}$ together any way you like; once you've gone out at least m steps, you can couple n blocks together very closely.

557. Exercise. Fill in the details. $\qquad\square$

Recapitulation of the proof.

The proof of the boldfaced theorem is spread out over this whole chapter, which might make it difficult for the reader to grasp the big picture. Here we give a brief recapitulation of the main ideas for quick reference.

558. B $\subset$ FB: This part was obvious.

559. FB $\subset$ IC: Every **FB** process is a factor of an independent process, hence a dbar limit of finite codings of an independent process. Finite codings of independent processes are approximable by their own period n independent concatenations, hence are **IC** according to Theorem 485.

560. IC $\subset$ EX: We proved in Theorem 501 that independent processes are completely extremal. Then we argued that an independent concatenation of length k is practically like an independent process on Λ^k. Finally, we used the fact that the class of extremal processes is closed in dbar (Theorem 519).

561. FD $\subset$ IC: We just observed that the e-startover process of a process is close to it in distribution and entropy and applied the fact that ϵ-startover processes are **IC** (Theorem 527).

562. EX $\subset$ VWB: This followed from Exercise 550 (most conditioned futures are nearly exponentially fat submeasures of the unconditioned future disjoint from a prechosen small set).

563. EX $\subset$ FD: We used the fact that Theorem 543 kicks in once most names are reasonable. We picked m big enough so that most names were reasonable, and so that exponentially fat submeasures of μ_m were close to μ_m in dbar. We then picked an approximating process in m-distribution and entropy. Both conditioned m-futures were usually fat submeasures of the unconditional, and had essentially the same unconditional, so could be coupled to each other.

564. VWB $\subset$ IC: We started with a very weak Bernoulli system and coupled its period n independent concatenation with the original, n-block by n-block.

$\square$

Here now are a couple of easy consequences of the boldfaced theorem.

565. Theorem. *All of the equivalent properties above are closed under dbar limits.*

Idea of proof. IC is closed under dbar limits. $\square$

566. Theorem. *All of the equivalent properties above are closed under the taking of factors.*

Idea of proof. IC is closed under finite code factors, and under dbar limits.

$\square$

567. Theorem. *Any member of the equivalent classes considered above is either trivial or has positive entropy.*

Idea of proof. Zero entropy means past determines future, easily contradicting **IC**. $\square$

6

Ornstein isomorphism theorem

In this chapter we prove the Ornstein isomorphism theorem, which states that any two finitely determined processes of the same entropy are isomorphic. In particular, since Bernoulli processes are finitely determined, this establishes that any two Bernoulli processes of equal entropy are isomorphic. (An earlier, related result of Sinai (1964) has, as a consequence, that any two Bernoulli processes of equal entropy are *weakly isomorphic*; i.e. each is a factor of the other. Both may be viewed as limiting, stationary versions of Shannon's (1948) noiseless coding theorem (1948).) Several (substantively different) proofs of Ornstein's theorem have been published; to the best of our knowledge, the one presented here adds to the variety. The closest match to our proof may be the proof of J. Kieffer (1984). Like Kieffer, we avoid the use of a marriage lemma.

568. Comment. In this chapter it is necessary to assume that our process is invertible (see, however Hoffman and Rudolph 2002, in which a condition is given for some non-invertible processes to be isomorphic to a one-sided Bernoulli shift). Also, though we restrict attention to the finite entropy case, the theorem has an infinite entropy version; see Ornstein (1970*b*).

6.1. Copying in distribution

In this section, we show that, given a process on a finite alphabet and a second system, we can construct a partition of the second system so that the second system with this partition is close in distribution to the first process. The reader should think of this as a very minor step in the direction of establishing an isomorphism. (It has to be minor, because you can do it for non-isomorphic systems!)

569. Definition. Let $(\Omega, \mathcal{A}, \mu)$ and $(Y, \mathcal{B}, \nu)$ be probability spaces and let P be a countable measurable partition of Ω. A *copy* of P on Y is an ordered pair (Q, π), where Q is a measurable partition of X and $\pi : P \to Q$ is a bijection (called the *copy map*) such that $\nu\big(\pi(p)\big) = \mu(p)$ for all $p \in P$.

570. Exercise. If $\pi : P \to Q$ is the copy map then π has a unique additive extension $\overline{\pi}$ (additive means satisfying $\overline{\pi}(A \cup B) = \overline{\pi}(A) \cup \overline{\pi}(B)$ whenever $A \cap B = \emptyset$) to the algebra generated by the cells of P.

571. Comment. We may suppress mention of the copy map π and say simply that Q is a copy of P.

Suppose one is given $n \in \mathbf{N}$, systems $(\Omega, \mathcal{A}, \mu, T)$ and $(Y, \mathcal{B}, \nu, S)$ and a finite measurable partition P on Ω. We will outline a technique for copying P to a partition Q on Y so that the (Q, S)-process has a distribution over n-names that is as close as desired in variation to that of the (P, T)-process. (We will assume that the space Y is non-atomic.)

572. Copying (P, T) to R.

Suppose we are given a Rohlin tower R of height N on Y. Let R' be a Rohlin tower on Ω of height N with an error set of the same size.

Now we partition R' into vertical columns according to the P-name of length N at the base. Partition R into the same number of vertical columns of the same size, and associate to each vertical column of R a unique column of R' having the same size. We now construct our copy Q of P on Y. The cells of Q will be denoted by $\{q_p : p \in P\}$ and $\pi(p) = q_p$ will be the copy map. Here is the construction: if C is a column of R and C' is the corresponding column of R' then each rung of C will lie in some $p \in P$. Just put the corresponding rung of C' into q_p. This handles the tower R. Extend to the error set in the obvious way. (Break the error set of R' into pieces according to P, make pieces of the same sizes in the error set of R and match them up.)

573. Definition. The procedure just outlined will be referred to as *copying (P, T) to R to get (Q, S)*.

574. Comment. If $n \in \mathbf{N}$, P is a partition on Ω and Q is a copy of P on Y, we can compare the distributions on n-names of the (P, T)-process and the (Q, S)-process by using the copy map π. The distributions of 1-names will coincide by the definition of a copy but the distributions of n-names for $n > 1$ can be much different. The reader should think about this until it's clear.

575. Definition. Let $n \in \mathbf{N}$. If (Q, S) is a copy of (P, T) and the two n-name distributions are less than ϵ apart in variation, we shall say that (Q, S) *n-resembles (P, T) up to ϵ*.

576. Comment. We may suppress mention of ϵ, saying simply that (Q, S) n-resembles (P, T). (Here it is assumed that ϵ is small.)

577. Exercise. Let $n \in \mathbf{N}$. Suppose that $N > \frac{n}{2\epsilon}$ and R is a Rohlin tower of height N on Y with error set less than $\frac{\epsilon}{2}$ in measure. Show that if (P, T) is copied to R to get (Q, S) then the distribution of n-names in (P, T) differs in variation from the distribution of n-names in (Q, S) by at most ϵ. *Hint: the n-names are the same except on the error set and on the uppermost n rungs of the towers.*

The previous exercise prompts the following definition.

578. Definition. Let $n \in \mathbf{N}$. If we copy (P, T) to R, where the error set of R has negligible measure and the height of R is much bigger than n, we will say we are *copying the n-distribution of* (P, T) *to* S.

579. Comment. Speaking informally, we may simply say that we *copy* (P, T) *to* R *to get* (Q, S), or, to be more long-winded, *copy* (P, T) *to* R *to get* Q *so that* (Q, S) *n-resembles* (P, T).

Let (Q, π) be a copy of P on Y, let P' be a refinement of P (all cells of P are a union of cells of P') and let (Q', π') be a copy of P'.

580. Definition. We say that (Q', π') *extends* (Q, π) (or simply that π' extends π) if $\overline{\pi'}$ agrees with π on P.

581. Exercise. Suppose P' refines P and (Q, π) is a copy of P on Y. Show that if Y is non-atomic then there is an extension (Q', π') of (Q, π).

The previous exercise shows that a copy can always be extended to a finer partition. The next example shows that you can't always do this in a way that preserves closeness of n-name distribution.

582. Example. Let $(\{a, b\}^{\mathbf{Z}}, \mu, T)$ be a system, where T is the shift and μ is concentrated uniformly on the four doubly infinite words $\cdots aaaaaa \cdots$, $\cdots ababab \cdots$, $\cdots bababa \cdots$ and $\cdots bbbbbb \cdots$ (μ assigns each of these words measure $\frac{1}{4}$). Now let $(\{h, t\}^{\mathbf{Z}}, \nu, S)$ be the standard $\frac{1}{2} - \frac{1}{2}$ Bernoulli system. Let X_i be the ith coordinate on the space of T and let Y_i be the ith coordinate on the space of S. Let $P = \big\{\{X_0 = a\}, \{X_0 = b\}\big\}$, $Q = \big\{\{Y_0 = h\}, \{Y_0 = t\}\big\}$ and $B = \big\{\{X_1 = a\}, \{X_1 = b\}\big\}$. Letting $\pi(\{X_0 = a\}) = \{Y_0 = h\}$ and $\pi(\{X_0 = b\}) = \{Y_0 = t\}$, Q is a copy of P.

583. Exercise. Verify that (Q, S) 2-looks (exactly) like (P, T), but that for no extension $(Q \vee C, \pi')$ of $(P \vee B, \pi)$ does $(Q \vee C, S)$ 2-look like $(P \vee B, T)$ up to $\frac{1}{4}$. *Hint: if $Y_0(y) \neq Y_2(y)$ then either y or Sy has a $(Q \vee C)$-name of length 2 whose translation to a $(P \vee B)$-name has probability zero.*

One might think that the previous example depends crucially on the fact that the space of T is atomic. This is not so, however.

584. Exercise. Show that the previous example still works if $(\{a, b\}^{\mathbf{Z}}, \mu, T)$ is replaced by its ϵ-startover process, where ϵ is small. (You can change $\frac{1}{4}$ to something slightly smaller.)

Copies made by copying to tall enough towers can, on the other hand, be extended while preserving closeness of n-name distribution: suppose you just copied (P, T) to a tower R having a small error set, where the height N of R is much bigger than n, to get (Q, S). Note that (Q, S) n-resembles (P, T). Now you are given another partition B on the space of T.

585. Exercise. Let π be the original copy map. Show that there is a partition C on the space of S such that $(Q \vee C)$ is a copy of $(P \vee B)$ by a copy map π' such that π' extends π, and that $(Q \vee C, S)$ n-resembles $(P \vee B, T)$. *Hint: each original column of R' splits into subcolumns according to B-name of length N. Split the corresponding columns of R into equal sized subcolumns, etc.*

586. Definition. When we do the above, we say we are *copying (B, T) to R to get C* so that $(Q \vee C, S)$ n-resembles $(P \vee B, T)$.

587. Theorem. *Let $\epsilon > 0$ and $n \in \mathbf{N}$. Let $N > n$ and suppose (Q, π) is a copy of P such that (Q, S) N-resembles (P, T) up to ϵ. Then for any measurable partition B on Ω, there is a measurable partition C on Y and a copy map $\pi' : P \vee B \to Q \vee C$ that extends π such that $(Q \vee C, S)$ n-looks like $(P \vee B, T)$ up to $\epsilon + \frac{n}{N}$.*

Idea of proof. Choose $\delta > 0$ such that the variation distance between the N-name distributions is less than $\epsilon - \delta$. Choose Rohlin towers R and R' on the spaces having error sets less than δ in measure and such that the bases are independent of the distributions of N-names for (P, T) and (Q, S), respectively. Split each tower into columns according to N-name. By shaving off and adding to the error set at most $\epsilon - \delta$ in measure from the columns of each tower, one can arrange that corresponding columns have the same measure. Now the tower on Y looks like it just had (P, T) copied to it, so you can proceed as in Exercise 585. n-names agree except possibly on the error set and top n rungs of the columns. $\square$

588. Corollary. *If (Q, S) is a perfect copy of (P, T) (exactly the same N distribution for all N), then for any n, ϵ and B you can copy B to get C so that $(Q \vee C, S)$ n-resembles $(P \vee B, T)$ up to ϵ.*

589. Definition. Let $(\Omega, \mathcal{A}, \mu)$ be a probability space. Suppose P and Q are measurable partitions of Ω and that $\pi : P \to Q$ is a bijection. Then we will write $P \cap_\pi Q = \bigcup_{p \in P} p \cap \pi(p)$ and $P \triangle_\pi Q = \Omega \setminus (P \cap_\pi Q)$. We also write $|P - Q|_\pi = \mu(P \triangle_\pi Q)$.

6.2. Coding

In this short subchapter we discuss approximation of factor maps by finite codes and restricting copies to a sub-σ-algebra. These are technical tools we will need later in the chapter.

Suppose π is a homomorphism from (P, T) to (Q, S). Then π can be approximated by a finite code. That is, there is some n such that the P-name $x_{-n}, x_{-n+1}, \ldots, x_{n-1}, x_n$ of a point x in the space of T determines the cell of Q containing $\pi(x)$ up to some small probability ϵ.[89] More precisely: for every $\epsilon > 0$, there is an n and a function f taking words of length $2n + 1$ $a_{-n} a_{-n+1} \cdots a_n$ to cells of Q such that

$$P\big(\pi(x)_0 = f(x_{-n} x_{-n+1} \cdots x_n)\big) > 1 - \epsilon.$$

590. Definition. In the above eventuality, we will say that (P, T) codes (Q, S) ϵ-well by time n. If $T = S$, we will say simply that (P, T) codes Q ϵ-well by time n. The function f is called an n-code from (P, T) to (S, Q).

591. Comment. In the language of processes, an n-code from $(X_i)_{i=-\infty}^{\infty}$ to $(Y_i)_{i=-\infty}^{\infty}$ is a function from located words $x_{-n} x_{-n+1} \cdots x_n$ on the alphabet of X to letters y_0 in the alphabet of Y. If $(Y_i)_{i=-\infty}^{\infty}$ is a factor of $(X_i)_{i=-\infty}^{\infty}$, we will say that *the X process codes the Y process ϵ-well by time n*, if $X_{-n} X_{-n+1} \cdots X_n$ determines Y_0 up to probability ϵ.

592. Comment. Suppose the X process codes the Y process ϵ-well by time n. Then for every $i \in \mathbf{Z}$, you can use the code f to predict with accuracy rate better than $1 - \epsilon$ what Y_i is on the basis of what $X_{i-n} X_{i-n+1} \cdots X_{i+n}$ is; indeed one has

$$P\big(Y_i = f(X_{i-n} X_{i-n+1} \cdots X_{i+n})\big) > 1 - \epsilon.$$

593. Exercise. Suppose that (P, T) codes Q ϵ-well by time n. If $|P - P_0| < \gamma$ then (P_0, T) codes Q $\big(\epsilon + (2n + 1)\gamma\big)$-well by time n.

594. Theorem. *Let $(X, \mathcal{A}, \mu, T)$ be a measure-preserving system, and suppose there are a sequence of measurable partitions (P_i), an increasing sequence of natural numbers (n_i) and a sequence of positive reals (ϵ_i) with $\sum_i \epsilon_i < \infty$, $|P_{i+1} - P_i| < \frac{\epsilon_i}{n_i}$ for all i, and such that for all i, (T, P_i) codes Q ϵ_i-well by time n_i. Then P_i converges to a measurable partition P such that (T, Q) is a factor of (T, P).*

---

[89] The reason for this is that $\pi^{-1}(Q)$ is a partition which, in the language of processes, can be approximated by a partition of cylinder sets.

Idea of proof. We need to show that the P-name determines the Q-name with probability 1. By the foregoing exercise, P_2 codes Q $4\epsilon_1$-well by time n_1. P_3 codes Q $(4\epsilon_1 + 3\epsilon_2)$-well by time n_1 and $4\epsilon_2$-well by time n_2. In general, P_m codes Q $(4\epsilon_l + 3\epsilon_{l+1} + \cdots + 3\epsilon_{m-1})$-well by time n_l. Hence the limit P should code Q $\left(4\sum_{i=l}^{\infty} \epsilon_i\right)$-well by time n_l.

595. Exercise. Complete the proof. $\qquad\qquad\qquad\qquad\qquad\qquad$ □

Now we discuss copying in a sub-σ-algebra.

596. Definition. Let $(X, \mathcal{A}, \mu, T)$ be a measure-preserving system and let P be a measurable partition that does not necessarily generate $\mathcal{A}$. The σ-algebra $\bigvee_{i=-\infty}^{\infty} T^i P$ will be called *the land of P*.

597. Comment. When copying, constructing partitions, etc. one can restrict everything to the land of P (that is, use only sets measurable with respect to the land of P), since by Exercise 61 the land of P gives rise to a Lebesgue space. Note in particular that if you copy a process to T in the land of P, the copy will be a factor of (P, T).

598. Example. Fix a system $(X, \mathcal{A}, \mu, T)$ and $n \in \mathbf{N}$. Suppose P, P_1 and P_2 are measurable partitions, and you have a copy P_3 of P_1 so that (P_3, T) n-looks exactly like (P_1, T). You would like to make a good copy P_4, in the land of P, of P_2 so that $(P_3 \vee P_4, T)$ n-resembles $(P_1 \vee P_2, T)$. You can do so as long as P_3 resides in the land of P (it doesn't matter whether P_1 and P_2 reside there). If P_3 does not reside in the land of P, there is no reason to think you can do it, because you use the (P_3, T)-process to construct P_4.

6.3. Capturing entropy: preparation

In the next section we will show how to copy a distribution while holding up the entropy, which is the most important technique in this chapter. We have a unique way of holding up the entropy, which is in many proofs done with a marriage lemma. In this section, we make the needed preparations.

599. Discussion. Let (P, T) be a process and let (Y, S) be a measure-preserving system. Copy the n distribution of (P, T) to S to get (Q, S) so that (Q, S) n-resembles (P, T). Question: is there anything we can say about the entropy of (Q, S) compared with that of (P, T)? To answer this, recall that (converting to the current notation) by Corollary 423 and Theorem 424 the entropy of (P, T) is the non-increasing limit of

$$\frac{1}{n} H \left(\bigvee_{i=1}^{n} T^{-i} P \right). \qquad\qquad (*)$$

This means that the n-distribution of (P, T) already gives an upper bound for the entropy of (P, T). Hence if (Q, S) has been selected to have approximately the same n-distribution as (P, T), and if n is large enough for ($*$) to be close to its limit, then (Q, S) could not possibly have (much) more entropy than (P, T).

However, the entropy of (Q, S) could conceivably be much less than the entropy of (P, T) (indeed, it could be zero), regardless of "how well" we copied the distribution. In short, a good copy of the n-distribution, for large n, guarantees a small enough entropy (up to ϵ), but does not guarantee a large enough entropy. If we want to guarantee that (Q, S) has approximately the same entropy as (P, T), we need to have some way of holding its entropy up. We are now going to develop techniques for copying distributions of processes (P, T) in such a way as to hold up the entropy of the copy (Q, S), thereby ensuring that (Q, S) and (P, T) have approximately the same entropy.

600. Definition. Let (P, T) be a process and let $n \in \mathbf{N}$. The (P, T, n) *entropy drop* is the number

$$\frac{1}{n} H \left(\bigvee_{i=1}^{n} T^{-i} P \right) - H(P, T).$$

601. Exercise. Fix a measure ν_n on words of size n. Prove that among all stationary processes whose distribution on n-names coincides with ν_n, the (unique) $(n-1)$-step Markov chain with this n-name distribution has minimum (P, T, n) entropy drop.

602. Theorem. *Let Q be a refinement of P. Then*

$$H(P) - \frac{1}{2}H(P \vee TP) \le H(Q) - \frac{1}{2}H(Q \vee TQ).$$

Proof. By Theorem 421, if A, B, and C are three partitions such that C refines B, then $H(A \vee C) - H(C) = H(A|C) \le H(A|B) = H(A \vee B) - H(B)$. Taking $A = TQ$, $B = P$ and $C = Q$, one gets

$$H(Q \vee TQ) - H(P \vee TQ) \le H(Q) - H(P).$$

Taking $A = P$, $B = TP$ and $C = TQ$, one gets

$$H(P \vee TQ) - H(P \vee TP) \le H(TQ) - H(TP) = H(Q) - H(P).$$

Add these two inequalities together.   □

603. Corollary. *Let Q be a refinement of P. Then*

$$\frac{1}{n} H\left(\bigvee_{i=1}^{n} T^i P\right) - \frac{1}{2n} H\left(\bigvee_{i=1}^{2n} T^i P\right) \leq \frac{1}{n} H\left(\bigvee_{i=1}^{n} T^i Q\right) - \frac{1}{2n} H\left(\bigvee_{i=1}^{2n} T^i Q\right).$$

Proof. Apply Theorem 602, replacing P by $\bigvee_{i=1}^{n} T^i P$, Q by $\bigvee_{i=1}^{n} T^i P$, and T by T^n. □

604. Corollary. *If Q is a refinement of P then for every $n \in \mathbf{N}$ the (P, T, n) entropy drop is less than or equal to the (Q, T, n) entropy drop.*

Idea of proof Use Corollary 603 together with the observation that the (P, T, n) entropy drop is equal to

$$\frac{1}{n} H\left(\bigvee_{i=1}^{n} T^{-i} P\right) - H(P, T)$$

$$= \lim_{k \to \infty} \left(\left(\frac{1}{n} H\left(\bigvee_{i=1}^{n} T^{-i} P\right) - \frac{1}{2n} H\left(\bigvee_{i=1}^{2n} T^{-i} P\right)\right)\right.$$

$$+ \left(\frac{1}{2n} H\left(\bigvee_{i=1}^{2n} T^{-i} P\right) - \frac{1}{4n} H\left(\bigvee_{i=1}^{4n} T^{-i} P\right)\right)$$

$$\left. + \cdots + \left(\frac{1}{2^k n} H\left(\bigvee_{i=1}^{2^k n} T^{-i} P\right) - \frac{1}{2^{k+1} n} H\left(\bigvee_{i=1}^{2^{k+1} n} T^{-i} P\right)\right)\right).$$

□

605. Theorem. *Let $(\Omega, \mathcal{A}, \mu, T)$ be a measure-preserving system, and let P be a measurable partition such that (P, T) is ergodic and has entropy H. If $\epsilon > 0$ then for all sufficiently large n and for all Rohlin towers R of height n and base B (having negligible error set), if one partitions B according to P-name of length n, thus creating P-name columns in R, then for some exceptional set E (a union of columns), each remaining column has measure between $2^{-n(H+\epsilon)}$ and $2^{-n(H-\epsilon)}$.*

Idea of proof. This theorem immediately follows if we know that the bad set of the Shannon–McMillan–Breiman theorem does not contain a large fraction of the base because the width of a column is exactly the size of a word in the base of that column. It's not easy to prove directly that the base has a small bad set, but we will now point out that there are nearby rungs with small bad sets. Pick γ and δ such that $0 < \delta \ll \epsilon$ (in particular we would like $(1 - \delta)(H - \delta) > H - \epsilon)$ and $0 < \gamma \ll \delta\epsilon$. By Shannon–McMillan–Breiman, there are $n \in \mathbf{N}$ and a bad set E of measure at most γ such that for all x outside of the bad set E and all $N > (1 - \delta)n$, the N-name of x has size between $2^{-N(H+\delta)}$ and $2^{-N(H-\delta)}$.

Let now R be a tower of height n with base B and negligible error set. Consider the rungs of the tower $B, TB, \ldots, T^{\lfloor \delta n \rfloor} B$. Not all of these rungs can intersect E in a set of measure more than $\frac{\epsilon}{2n} \approx \frac{\epsilon}{2}\mu(B)$ (there isn't enough of E to go around), so some one of these rungs r must intersect the bad set E in a set of measure less than $\frac{\epsilon}{2n}$. The existence of r establishes that all but $\frac{\epsilon}{2}$ of the columns of R are no larger than $2^{-n(H-\epsilon)}$ in measure.[90] A similar argument with the sets $B, T^{-1}B, \ldots, T^{-\lfloor \delta n \rfloor} B$ establishes the required estimate from below.   □

606. Abbreviation. Since the bad set of Shannon–McMillan–Breiman is small, there are nearby rungs above and below the base intersecting the bad set only slightly.

607. Exercise. *Young children's puzzle theorem.* Let P be an ordered partition with m cells and let Q be a partition whose sets are all smaller than ϵ in measure. Then there is a partition P' such that (a) Q is finer than P', and (b) $v(P, P') < m\epsilon$.

608. Exercise. Show that it is possible to decrease the entropy of a partition arbitrarily much by lumping many tiny cells together, even when the resulting lump has arbitrarily small measure.[91] *Hint: use the conditioning property.*

609. Lemma. *Let P be a measurable partition with m pieces, let $n \in \mathbf{N}$, and let B be a set of n-names in the (P, T) process. If $\mu(B) \ll \frac{1}{\log m}$, then lumping the members of B together[92] will not significantly alter the value of $\frac{1}{n} H\left(\bigvee_{i=0}^{n-1} T^{-i} P \right)$.*

Idea of proof. The number of possible names of length n is at most m^n. The normalized measure on B therefore has at most m^n pieces. The largest its entropy can be is if all those sets have the same entropy, in which case it is $n \log m$. By the conditioning property, therefore, the contribution of B to $H\left(\bigvee_{i=0}^{n-1} T^{-i} P \right)$ is at most $\mu(B)n \log m$.   □

---

[90] Taking $N = n - j$, where $r = T^j B$, one sees that any point in r whose N-name is not bigger than $2^{-N(H-\delta)}$ lies in a column of the tower having size at most $2^{-N(H-\delta)} < 2^{-n(H-\epsilon)}$. Moreover any point in $r \setminus E$ is such a point.

[91] To be precise: for any $M > 0$ and any $\epsilon > 0$ there is a measurable partition P and a set Σ of cells of P such that $\mu(\bigcup \Sigma) < \epsilon$ such that if P' is the partition that results from lumping the members of Σ into a single cell, then $H(P') < H(P) - M$.

[92] We identify an n-name with that atom of $\bigvee_{i=0}^{n-1} T^{-i} P$ consisting of those points having that n-name.

610. Comment. In the preceding lemma, B doesn't actually have to be a union of n-names. If B is just some measurable set satisfying $\mu(B) \ll \frac{1}{\log m}$, and E denotes $\{B, B^c\} \vee \bigvee_{i=0}^{n-1} T^{-i} P$, then lumping all of the cells of E contained in B together will still not significantly affect $\frac{1}{n} H(E)$.

Copying a distribution while holding up the entropy.

611. Theorem. *Let (P, T) be a process, $n \in \mathbf{N}$, and let $(Y, \mathcal{B}, \nu, S)$ be a measure-preserving system with $H(S) > H(P, T)$. For any $\delta > 0$ there is a finite measurable partition Q on Y so that the n-distributions and entropies of (P, T) and (Q, S) are within δ of one another.*

Idea of proof. For the purposes of this proof $x \approx y$ will mean that $|x - y| \ll \delta$. Put $H = H(P, T)$, $H' = H(S)$ and let $\epsilon \ll \min\{H(S) - H(P, T), \delta\}$. Our first observation is that it doesn't hurt to assume that n is larger; that is, to copy the n'-distribution, where $n' > n$, is to copy the n-distribution as well. Accordingly, we may choose a generator G for S and assume that n is large enough that $H(S) \approx \frac{1}{n} H\left(\bigvee_{i=0}^{n-1} S^i G\right)$. That is, we assume that the (G, S, n)-entropy drop is small. Also we assume that n is large enough that $\frac{1}{n} H\left(\bigvee_{i=0}^{n-1} T^i P\right) \approx H(P, T)$ and $\frac{\log n}{n} \ll \epsilon$.

Now we proceed. Among other things we are going to be copying the n-distribution, which we already know how to do: let $N = (n + 1)n$, and choose N-towers R for T and R' for S. Our goal is to copy P from R to R' carefully, so that we keep our entropy.

By Theorem 605, if N is large enough then only a small measure of (P, T, R)-columns have measure smaller than $2^{-N(H+\epsilon)}$. Alter P by lumping those "bad" columns together into one big column with one name (any given name you choose); by Lemma 609 that won't change either the entropy or distribution of (P, T) very much. Now you have at most $2^{N(H+\epsilon)}$ columns.

Consider the (S, G, R')-columns. If N is large enough, all but a very small measure of those columns have measure at most $2^{-N(H'-\epsilon)}$. This is much less than the reciprocal of the number of (P, T, R)-columns. Therefore, by the young children's puzzle theorem, we can copy (P, T) to R' to get Q in such a way that each (S, Q, R')-column is a union of (S, G, R')-columns.[93] Put the whole error set into a single cell of Q.

Denote the rungs of R' by $B_0, B_1, \ldots, B_{(n+1)n-1}$. Let P_G be the partition whose pieces are the rungs of the G-columns of R', excepting the uppermost

---

[93] The copy will be less precise than it normally would be; there will be a small error which comes from the puzzle theorem and another from the big (S, G, R')-columns. These errors do not add appreciably, however, to the distribution errors one already gets from the error sets of the towers and from upper rung effects.

$n - 1$ rungs of the tower, and a big chunk consisting of the top $n - 1$ rungs of the tower and the error set. Similarly let P_Q be the partition whose pieces are the rungs of the Q-columns of R', excepting the uppermost $n - 1$ rungs of the tower, and the same big chunk. Note that P_G is finer than P_Q. Note also that P_G practically generates.[94] We claim that the (P_G, S, n)-entropy drop is approximately equal to the (G, S, n)-entropy drop (which is small by assumption).

612. Exercise. Prove the claim by justifying the following.

(a) $H(G, S) \geq H(P_G, S)$. (Obvious since G generates.)
(b) If the error set of R' is small then $H(P_G, S) \geq H(G, S) - \epsilon$. *Hint: use the prior footnote, smallness of the error set of R', and Exercise 429.*
(c) Write $\overline{B} = \{B_0, B_n, B_{2n}, \ldots, B_{n^2}, (B_0 \cup B_n \cup B_{2n} \cup \cdots \cup B_{n^2})^c\}$. Then $\bigvee_{i=0}^{n-1} S^i(G \vee \overline{B})$ is a refinement of $\bigvee_{i=0}^{n-1} S^i P_G$.
(d) $\left| \frac{1}{n} H(\bigvee_{i=0}^{n-1} S^i(G \vee \overline{B})) - \frac{1}{n} H(\bigvee_{i=0}^{n-1} S^i G) \right| < \epsilon$. *Solution: consider that*

$$\frac{1}{n} H \left(\bigvee_{i=0}^{n-1} S^i(G \vee \overline{B}) \right) = \frac{1}{n} H \left(\left(\bigvee_{i=0}^{n-1} S^i G \right) \vee \left(\bigvee_{i=0}^{n-1} S^i \overline{B} \right) \right)$$

$$\leq \frac{1}{n} H \left(\bigvee_{i=0}^{n-1} S^i G \right) + \frac{1}{n} H \left(\bigvee_{i=0}^{n-1} S^i \overline{B} \right)$$

$$\leq \frac{1}{n} H \left(\bigvee_{i=0}^{n-1} S^i G \right) + H(\overline{B}).$$

Now use the fact that $H(\overline{B}) < \epsilon$, n large. ●

Next, we claim that the (P_Q, S, n)-entropy drop is approximately equal to the (Q, S, n)-entropy drop.

613. Exercise. Prove this claim by justifying the following.

(a) $H(P_Q, S) \geq H(Q, S)$. *Hint: $\bigvee_{i=0}^{n-1} S^i P_Q$ is a refinement of Q.*
(b) $\bigvee_{i=0}^{n-1} S^i(Q \vee \overline{B})$ is a refinement of $\bigvee_{i=0}^{n-1} S^i P_Q$.
(c) For $m \geq n$, $\left| \frac{1}{m} H(\bigvee_{i=0}^{m-1} S^i(Q \vee \overline{B})) - \frac{1}{m} H(\bigvee_{i=0}^{m-1} S^i Q) \right| \leq H(\overline{B}) < \epsilon$.
(d) $H(Q, S) \geq H(P_Q, S) - \epsilon$. *Hint: let $m \to \infty$ in (c); use (b).* ●

Now just use the fact that (since P_G refines P_Q) the (P_Q, S, n)-entropy drop is no more than the (P_G, S, n)-entropy drop to see that the (Q, S, n)-entropy drop must be small. In other words, $H(Q, S) \approx \frac{1}{n} H(\bigvee_{i=0}^{n-1} S^i Q) \approx \frac{1}{n} H(\bigvee_{i=0}^{n-1} T^i P) \approx H(P, T)$.

---

[94] Not quite, but close. The σ-algebra generated by $\bigvee_{i=0}^{n-1} S^i P_G$ will contain every member of G's intersection with R', however.

614. Exercise. Fill in the details to complete the proof. □

615. Abbreviation. To copy an n-distribution and keep entropy, assume WLOG $n \gg 1$ and copy one $N = (n + 1)n$ tower onto another as usual. First put G columns on the range tower where G is a generator and then compose copy columns out of G columns by fudging columns to make the puzzle theorem applicable. Intersect columns with rungs not near the top to get P_Q and P_G, use that the (P_Q, S, n)-entropy drop is no greater than the (P_G, S, n)-entropy drop.

616. Corollary. *Theorem 611 remains valid if* $H(S) = H(P, T)$.

Idea of proof. Don't copy (P, T) itself, but something close to it. That is, choose a partition P_0 such that $H(P_0, T)$ is slightly less than $H(P, T)$ but such that (P_0, T) is close to (P, T) in distribution. Then use the proof of Theorem 611 to copy (P_0, T). One way to get P_0 is to form a very tall Rohlin tower and lump a small set of columns together. A small lumping won't affect the n-distribution much, nor will it do much damage to entropy. Nevertheless to see that one can lower entropy in this way, at least somewhat, just pick a big N such that $\frac{1}{N} H\left(\bigvee_{i=1}^{N} T^i P \right)$ is very close to $H(P, T)$, and choose your tower of height N with very small error set. Forming P_0 by lumping a small set of columns together will be enough to get the entropy of the partition into rungs of P_0-columns low enough to force $H(P_0, T) < H(P, T)$. □

617. Exercise. Let $(X, \mathcal{A}, \mu, T)$ and $(Y, \mathcal{B}, \nu, S)$ be measure-preserving systems. Let P and Q be finite measurable partitions on X, let q be a finite measurable partition on Y and assume that (Q, T) is isomorphic to (q, S). Suppose that $H(S) \geq H(P \vee Q, T)$. For any $n \in \mathbf{N}$ and $\delta > 0$ there is a finite measurable partition p on Y so that the n-distributions and entropies of $(P \vee Q, T)$ and $(p \vee q, S)$ are within δ of one another. *Hint: adapt the proof of Theorem 611; incorporate ideas from Corollary 588 and Corollary 616.*

6.4. Tweaking a copy to get a better copy

What makes the proof of the Ornstein isomorphism theorem difficult is that you can't just construct an isomorphism in one go; you have to make better and better copies that converge to an isomorphism. When you do this, you have to make sure you can make a much better copy than the one you've got by making a very small change. In this section, we prove two theorems (we need them both) that allow one to do that, but first we need a technical lemma and a new puzzle theorem.

618. Lemma. *Fix* $a, \epsilon > 0$ *with* $a + \epsilon < \frac{1}{2}$. *For sufficiently large* n, *the number of words in* $\{0, 1\}^n$ *having less than an* $1s$ *is not more than* $2^{-n(a+\epsilon)\log(a+\epsilon)-n(1-a-\epsilon)\log(1-a-e)}$.

Idea of proof. One can of course prove this using Stirling's formula but we prefer this proof. Consider a random independent stationary process (X_i) on the alphabet $\{0, 1\}$ where $P(X_1) = a + \epsilon$. The entropy of this process is $-(a + \epsilon)\log(a + \epsilon) - (1 - a - \epsilon)\log(1 - a - e)$, so the typical reasonable word of length n has probability roughly $2^{n(a+\epsilon)\log(a+\epsilon)+n(1-a-\epsilon)\log(1-a-e)}$. Now simply observe that, since $a + \epsilon < \frac{1}{2}$, words having less than an 1s have higher probability than reasonable words. $\qquad\qquad\square$

619. Theorem. *Let* Λ *be a finite alphabet and fix* a, ϵ *with* $a + \epsilon < \frac{1}{|\Lambda|}$. *If* n *is sufficiently large and* $w \in \Lambda^n$ *is a word, then the number of words* $v \in \Lambda^n$ *such that* $d(w, v) < a$ *is not more than* $2^{-n(a+\epsilon)\log(a+\epsilon)-n(1-a-\epsilon)\log(1-a-e)+n(a+\epsilon)\log(|\Lambda|-1)}$.

Idea of proof. Don't use the previous lemma directly. Use the idea of its proof. First describe a set that has the same number of elements as the given set. $\quad\square$

620. Definition. Let Λ be a finite alphabet. We put

$$h(d) = h_\Lambda(d) = -d\log(d) - (1 - d)\log(1 - d) + d\log(|\Lambda| - 1).$$

621. Exercise. *Mature children's puzzle theorem.* Let M and P be two partitions such that $M \vee P$ contains m sets. Let M' be a partition with the same number of cells as M. Let Q be a partition finer than M' all of whose sets are of measure at most ϵ. Then there is a partition P' such that Q is finer than $M' \vee P'$ and the variation distance between the distributions of $M \vee P$ and $M' \vee P'$ is less than $m\epsilon$ plus the variation distance between the distributions of M and M'.

622. Discussion. Let $(X_i)_{i=-\infty}^\infty$ and $(Y_i)_{i=-\infty}^\infty$ be two processes on the same alphabet Λ. Let c be a coupling between them. Recall that $(X_i)_{i=-\infty}^\infty$ and $(Y_i)_{i=-\infty}^\infty$ can be identified with measures μ and ν on $\Lambda^{\mathbf{Z}}$, and c is a measure on the product space $\Lambda^{\mathbf{Z}} \times \Lambda^{\mathbf{Z}}$ whose marginals are μ and ν, respectively. Moreover, for fixed n we can look at the measures μ_n and ν_n that μ and ν induce on words of length n; c_n, the measure that c induces on pairs of words (v, w) of length n, couples these.

We now consider a completely new Lebesgue space having a Rohlin tower of height n. (You don't even have to assume there to be any error set; the entire space consists of n sets having measure $\frac{1}{n}$ each.) Break the tower into columns so that for every ordered pair $(v, w) \in \Lambda^n \times \Lambda^n$ there is an associated column whose measure is $c_n(v, w)$.

We now construct a partition of our new space into cells indexed by $\Lambda \times \Lambda$ in the obvious way: we regard the mth rung of the column associated with (v, w) to be in the cell $C_{(\lambda_1, \lambda_2)}$ indexed by (λ_1, λ_2) if the mth letter of v is λ_1 and the mth letter of w is λ_2.

623. Definition. A tower as constructed above, endowed with a partition indexed by $\Lambda \times \Lambda$, is called a *picture of the coupling between* $(X_i)_{i=1}^{n}$ *and* $(Y_i)_{i=1}^{n}$.

624. Exercise. Show that $\bigcup_{\lambda_1 \neq \lambda_2} C_{(\lambda_1, \lambda_2)}$ has measure equal to the expected mean Hamming distance achieved by the coupling.

625. Theorem. *Let* (P, T) *be a process, let* $n \in \mathbf{N}$ *and let* $(Y, \mathcal{B}, \nu, S)$ *be a measure-preserving system. Let* Q *be a partition on* Y *whose cells are in one-to-one correspondence with those of* P[95] *and let* $d = \overline{d}((P, T), (Q, S))$. *If* $H(S) > H(Q, S) + h(d^{\frac{1}{2}})$, *then for every* $\delta > 0$ *there exists a partition* Q_1 *on* Y *such that*

(1) Q_1 differs from Q by at most $3d^{\frac{1}{2}}$ in variation;
(2) $|H(P, T) - H(Q_1, S)| < \delta$; and
(3) (P, T) and (Q_1, S) have n-distributions differing by at most δ (in variation).

Proof. As in the proof of Theorem 611, we let G be a generator for S and assume without loss of generality (increasing n if necessary) that $\frac{1}{n} \ll \delta$ and that both the (G, S, n)-entropy drop and the (P, T, n)-entropy drop are $\ll \delta$. Moreover, we assume that for all $N > n$, the measure of unreasonable N-names for the (Q, S) process is $\ll \delta$. Fix $N = (n+1)n$. In the proof to follow, we will verify conclusions (1) and (3), leaving conclusion (2) as an exercise (the reader may wish to consult the proof of Theorem 611).

There is a coupling of the N-distribution of (P, T) with the N-distribution of (Q, S) which achieves an expected mean Hamming distance less than d. Let R'' be a picture of this coupling. Denote by P'' and Q'' the partitions on R'' corresponding to the first and second coordinates, respectively. For the moment ignore P'' and just focus on the Q'' columns. In preparation for copying, lump together all unreasonably small Q'' columns (totally ignoring what P'' looks like on those columns) as we did in the proof of Theorem 611. By assumption, this lump has some negligible size $\ll \delta$.

Now we will add some more $(P'' \vee Q'')$-columns to the lumped column. Consider a $(P'' \vee Q'')$-column in R'' to be "bad" if the proportion of its

---

[95] Given a one-to-one correspondence of this kind, (P, T) and (Q, S) can be regarded as stationary processes on the same alphabet.

rungs in which the first and second coordinates disagree exceeds $d^{\frac{1}{2}}$ (Our problem is that the number of bad columns may be too big for us to apply the mature children's puzzle theorem.) Before adding these columns to the lump, the lump was negligibly small. Now since d is the Hamming distance achieved by the coupling, the size of the lumped column is at most something like $d^{\frac{1}{2}}$.

On the lumped column, we erase all knowledge of P'' so that we no longer have $(P'' \vee Q'')$-columns, but rather only Q''-columns. Now we count columns. The number of unlumped Q''-columns can't be very much bigger than $2^{NH(Q,S)}$. By Theorem 619, the number of unlumped $(P'' \vee Q'')$-columns inside of an unlumped Q''-column can't be much bigger than $2^{Nh(d^{\frac{1}{2}})}$. Hence, the total number of unlumped $(P'' \vee Q'')$-columns can't be much greater than $2^{NH(Q,S)+Nh(d^{\frac{1}{2}})}$.

Let R be a tower of height N on Y having error set with measure $\ll \delta$. Since $H(S) > H(Q, S) + h(d^{\frac{1}{2}})$, R has something like $2^{NH(Q,S)}$ reasonable G-columns; this is enough to apply the puzzle theorems. Now we copy P'' to get Q_1. In copying the unlumped $(P'' \vee Q'')$-columns, use the mature children's puzzle theorem to get G finer than Q_1 and to get $Q_1 \vee Q$ to look like $P'' \vee Q''$. On the lumped column, use the young children's puzzle theorem to get G finer than Q_1 and to get Q_1 to look like P''.

Now, conditioned on being in an unlumped $(P'' \vee Q'')$-column, the probability of being in "different cells" (actually non-correlated cells) of P'' and Q'' is no more than $d^{\frac{1}{2}}$. Since the measure of the lumped column is also no more than $d^{\frac{1}{2}}$, and as the error coming from the puzzle theorems is not worse than the measure of a reasonable G-column times the number of unlumped $(P'' \vee Q'')$-columns, which is negligible, one gets $v(Q_1, Q) < 3d^{\frac{1}{2}}$. This gives (1). Finally, since the error coming from the puzzle theorems is negligible, the initial lump (where one erased knowledge of Q'') had size much less than δ, and since $n \gg \frac{1}{\delta}$, the variation distance between the distributions of n-names for P and Q_1 is small ($< \delta$).

626. Exercise. Verify (2) by mimicking the proof of Theorem 611. $\square$

627. Comment. Theorem 625 will be used, but it won't be quite enough. We will also need that in a special case, where we have a better picture of the coupling, we can copy more efficiently, and with a much lower entropy ceiling. We now proceed to develop this special case, in which (Q, S) is a factor of (P, T).

628. Definition. Let $(Y, \mathcal{B}, \nu)$ be a factor of $(X, \mathcal{A}, \mu)$ and let $\pi : X \to Y$ be the factor map. The *coupling induced by* π is the unique coupling of μ and ν

supported on the graph of π. That is, such that $c\big(\{(x, \pi(x)) : x \in A\}\big) = \mu(A)$ for every $A \in \mathcal{A}$.

629. Exercise. Show that there is such a coupling, and that it is unique.

630. Exercise. Let $(Y_i)_{i=-\infty}^{\infty}$ be a factor of $(X_i)_{i=-\infty}^{\infty}$, where these are processes on a finite alphabet, let π be the factor map, and let γ be the coupling corresponding to π. Let c_N be the measure on pairs of words of length N induced by γ. Recall that c_N couples μ_N and ν_N. Let $H = H\big((x_i)_{i=-\infty}^{\infty}\big)$. Let $\eta > 0$ be arbitrarily small. Show that, when N is large, upon exclusion of an error set having measure less than η, c_N is supported on no more than $2^{N(H+\eta)}$ pairs of names. *Hint: for big enough n, $H\big(Y_0 | \bigvee_{i=-n}^{n} X_i\big) < \frac{\eta}{2}$. Take $N \gg n$.*

Here now is the special version of Theorem 625 alluded to above.

631. Theorem. *Let (P, T) be a process, let $n \in \mathbf{N}$ and let $(Y, \mathcal{B}, \nu, S)$ be a measure-preserving system. Let Q be a partition on Y whose cells are in one-to-one correspondence with those of P and let $d = \overline{d}((P, T), (Q, S))$. If (Q, S) is a factor of (P, T) and $H(S) > H(P, T)$, then for every $\delta > 0$ there exists a partition Q_1 on Y such that*

(1) Q_1 differs from Q by at most 2d in variation;
(2) $|H(P, T) - H(Q_1, S)| < \delta$; and
(3) (P, T) and (Q_1, S) have n-distributions differing by at most δ (in variation).

Idea of proof. Basically the same proof as that of Theorem 625, only easier. There are no big lumps to worry about here, just a small one coming from Exercise 630. (The puzzle theorems apply without having to eliminate columns on which the first and second coordinates disagree by some given amount.) □

6.5. Sinai's theorem

In this section we prove Sinai's theorem, which states that every measure-preserving system has a finitely determined factor of full entropy. Then, we prove a stronger form of Sinai's theorem that we need for the proof of the Ornstein isomorphism theorem in the next section. Before doing any of that, we need a technical result stating that you can exhaust an arbitrary system by a chain of factors of increasing entropy satisfying some additional requirements.

The following exercise is intended to give a flavor of the exhaustion theorem that follows it, and may be skipped if desired.

632. Exercise. Let $(X_i)_{i=-\infty}^{\infty}$ be an independent, $(\frac{1}{2}, \frac{1}{2})$ process on the alphabet $\{a, b\}$. Now define a stationary process $(Y_i)_{i=-\infty}^{\infty}$ on the same space by the rules $Y_i = d$ if $X_i = b$ or $X_i X_{i+1} X_{i+2} X_{i+3} X_{i+4} = aaaaa$, and $Y_i = c$ otherwise.

(a) $(Y_i)_{i=-\infty}^{\infty}$ is a factor of $(X_i)_{i=-\infty}^{\infty}$.

(b) $H\big((Y_i)_{i=-\infty}^{\infty}\big) < H\big((X_i)_{i=-\infty}^{\infty}\big)$. *Hint: both the* baaaaab *and* bbaaaab *map to* ddccccd *on the same indices and have equal effect on other indices.*

(c) Define a process $(Z_i)_{i=-\infty}^{\infty}$ by $Z_i = f$ if $X_i = b$ or $X_i X_{i+1} \cdots X_{i+24} = aa \cdots a$. Show that $(Y_i)_{i=-\infty}^{\infty}$ is a factor of $(Z_i)_{i=-\infty}^{\infty}$. *Hint: first show that* ef *always comes from* ab, *then show that* $Z_i Z_{i+1} Z_{i+2} Z_{i+3} Z_{i+4}$ *determines* Y_i.

(d) Choose a large n and let γ be the coupling on pairs of words of length n induced by the factor map mapping $(Z_i)_{i=-\infty}^{\infty}$ to $(Y_i)_{i=-\infty}^{\infty}$. Show that the mean Hamming distance achieved by γ is $< 2^{-5}$ when n is large enough. *Hint:* $f \to d$ *and* $e \to c$ *are matches.* $f \to c$ *never happens, so* $e \to d$ *is the only mismatch and this happens* $< 2^{-5}$ *of the time.*

633. Theorem. *Let (P, T) be an ergodic positive entropy process on a finite alphabet. There exists a sequence of partitions P_i, each of whose cells are in one-to-one correspondence with the cells of P, such that:*

(a) *(P_i, T) is a factor of (P_{i+1}, T) (we denote the factor map by π_i);*

(b) *$\sum_i d_i < 1$, where d_i is the mean Hamming distance achieved by the coupling corresponding to π_i;*

(c) *$|P - P_i| \to 0$ as $i \to \infty$; and*

(d) *$H(P_i, T) < H(P, T)$ for all i.*

Idea of proof. To start, we want to choose P_1. Pick a big n_1 and start with the partition into P, n_1 names. Choose a name $a_0 a_1 \cdots a_{n_1-1}$. The set of points having this name lies in a_0. Relabel the points having that name to a different letter (some letter already in your alphabet). When you do this, entropy can only go down. If it does, stop. Otherwise, do it again. As soon as entropy goes down, stop. P_1 is the partition into the new labels. Since you can figure out which cell of P_1 you are in by looking at the infinite P-name, (P_1, T) is a factor of (P, T).

Now pick a much bigger n_2 and consider the partition into P, n_2 names. Make sure this partition is into sets so tiny that the entropy drop you can get by labeling any name with a different initial letter won't drop the entropy more than $H(P, T) - H(P_1, T)$. Now do what you did before. Pick a name $a_0 a_1 \cdots a_{n_2-1}$ and relabel the initial letter of all the points having that name. However, when you do this, only relabel names $a_0 a_1 \cdots a_{n_2-1}$ for which you already labeled the initial string $a_0 a_1 \cdots a_{n_1-1}$ in the first part of the proof, and relabel them in the same way as you did before. Keep lumping only long enough to make entropy go down a bit. That's P_2.

634. Exercise. Show that (P_1, T) is a factor of (P_2, T) is a factor of (P, T). Argue how to continue in this fashion and verify the remaining claims.   □

635. Notation. We will write $(Q, S) \approx (P, T)$ when the n-distributions of (Q, S) and (P, T) coincide for all n. Of course, by Exercise 145, this implies that (Q, S) and (P, T) are isomorphic, so we will interpret $\approx$ thusly.

636. Theorem. *(Sinai's theorem 1964.) Let $(Y, \mathcal{B}, \nu, S)$ be an ergodic system and let (P, T) be a finitely determined process. If $H(P, T) \leq H(S)$ then (P, T) is a factor of S.*

Idea of proof. The proof will be completed in a series of steps.

(1) Choose partitions P_i as in the conclusion of Theorem 633. Note that each (P_i, T) is finitely determined, as the FD property is closed under the taking of factors.

(2) By Theorem 611, choose a measurable partition P_{01} on Y such that (P_{01}, S) approximates (P_1, T) in distribution and entropy (hence in dbar).

(3) Use Theorem 625 to alter P_{01} slightly to get a new partition P_{02} such that (P_{02}, S) is a much better approximation to (P_1, T).[96]

(4) Continue in this fashion: alter P_{02} slightly to get a new partition P_{03} such that (P_{03}, S) approximates (P_1, T) better still, etc.[97] Make the approximations converge fast enough[98] that the sequence P_{0i} converges in variation to a partition P_{10} with $(P_{10}, S) \approx (P_1, S)$.

(5) Use Theorem 631 to alter P_{10} slightly to get a new partition P_{11} such that (P_{11}, S) is a good approximation of (P_2, T).

(6) Emulate steps (3)–(4), using Theorem 625 to make alterations to get partitions P_{12}, P_{13}, etc., converging to a partition P_{20} with $(P_{20}, S) \approx (P_2, S)$.[99]

---

[96] Theorem 625 requires a certain amount of excess entropy, relative to the dbar distance between the partition you are altering and the one you are approximating. One must therefore make sure in step 2 that (P_{01}, S) approximates (P_1, T) so closely in distribution and entropy that it will be also be close enough in dbar (recall (P_1, T) is finitely determined) to be able to do step (3). Similar remarks apply throughout the proof.

[97] Approximations in n-distribution, for larger and larger values of n.

[98] One needs to alter P_{0i} by at most $3d^{\frac{1}{2}}$, where $d = \overline{d}((P_{0i}, S), P_1, T))$. Since (P_1, T) is finitely determined, d can be made as small as required in the previous step, when P_{0i} was chosen to approximate (P_1, S) in distribution and entropy. In order that the sequence stabilize, one needs to make sure that the sizes of the alterations sum to something small.

[99] These alterations should be summably tiny (P_{20} is itself a very slight alteration of P_{10}).

(7) Follow the same recipe to get P_{30}, P_{40}, etc. converging[100] to some P_* such that $(P_*, S) \approx (P, T)$. $\qquad\qquad\qquad\qquad\qquad\qquad\qquad\qquad$ □

637. Comment. Notice that if $H(P, T) < H(S)$, one doesn't need to use Theorems 631 and 633.

638. Discussion. We now wish to develop a special case of Sinai's theorem, with an extra hypothesis and a stronger conclusion. Suppose (P, T) and (Q, S) are processes, and the cells of P are in one-to-one correspondence with the cells of Q. Let $\overline{d}\big((P, T), (Q, S)\big) = \epsilon$. Suppose now we have an alphabet Λ and a finite code f taking P-words of length $2n + 1$ to Λ. Letting $(x_i)_{i=-\infty}^{\infty}$ be the P-name of a point x, the map π sending x to $(a_i)_{i=-\infty}^{\infty}$, where $a_i = f(x_{i-n}x_{i-n+1}\cdots x_{i+n})$, determines a factor $(X_i)_{i=-\infty}^{\infty}$ of (P, T). Let f' be (essentially the same code) the function taking corresponding Q-words of length n to Λ and use f' to form a factor $(Y_i)_{i=-\infty}^{\infty}$ of (Q, S) in the same way. (Call the factor map π'.)

639. Exercise. Show that $\overline{d}\big((X_i)_{i=-\infty}^{\infty}, (Y_i)_{i=-\infty}^{\infty}\big) < (2n+1)\epsilon$. *Hint: a good dbar coupling of the original processes induces a good dbar coupling of the images, where k errors on terms from $-(M + n)$ to $(M + n)$ induce at most $(2n + 1)k$ errors on terms from $-M$ to M.*

640. Exercise. Suppose you used the same alphabets for the factors as for the original processes themselves. (That is, there is a one-to-one correspondence.) Then it makes sense to ask whether a given point when moved by the factor map π ends up in the corresponding cell in the factor space. Let γ be the probability that this doesn't happen; that is, the probability that x winds up in a different cell.[101] Show that the probability that a random point in the space of the (Q, S) process winds up in a non-corresponding cell in its factor space is at most $(2n + 1)\epsilon + \gamma$. *Hint: as before, induce a coupling from a coupling. If the $-n$ to n terms of (P, T) and (Q, S) coincide and the 0 coordinate of (P, T) and its factor coincide then the 0 coordinate of (Q, S) and its factor will coincide.*

641. Theorem. *Sinai's theorem, strong form of a special case.* For every $\epsilon > 0$ there exist $\delta > 0$ and m such that if (Q, S) is an ergodic process and (P, T) is a finitely determined process, where the cells of P are in one-to-one correspondence with the cells of Q, with $H(Q, S) - \delta < H(P, T) \leq H(Q, S)$ and such

---

[100] Of course, to guarantee convergence one has to provide that the successive alterations are summable in variation.

[101] This is, of course, $1 - P\big((\pi(x))_0 = g(c(x))\big)$, where $c(x)$ denotes the cell of P that x lies in and g is the bijection taking the cells of P to Λ.

that the variation distance of the m-distributions of (Q, S) and (P, T) differ by at most δ, then there is a partition P_* on the space of Q with $|Q - P_*| < \epsilon$ and such that $(P_*, S) \approx (P, T)$.

Idea of proof. Choose the P_i as in the proof of Sinai's theorem. We may assume that $\sum_i |P_{i+1} - P_i| \ll \epsilon$. Notice in the proof of Sinai's theorem, once P_{01} is in a position to be chosen, the amount it has to be altered to get to P_* is at most $2 \sum_i \overline{d}\big((P_{i+1}, T), (P_i, T)\big)$ plus a quantity that can be made as small as desired. Therefore, it suffices to show that we may choose P_{01} close to Q, modulo the restriction that $\overline{d}\big((P_{01}, S), (P_1, T)\big)$ be small enough to perform the future steps. Here are some more details.

(1) Once we choose P_{01} with (P_{01}, S) close to (P_1, T), we don't want to have to move P_{01} more than $\frac{\epsilon}{2}$ to get to P_*. There is some $\alpha \ll \epsilon$ such that if $\overline{d}((P_{01}, S), (P_1, T))) < \alpha$, then this will be the case.[102] All we must do therefore is choose P_{01} such that $|P_{01} - Q| < \frac{\epsilon}{2}$ and $\overline{d}((P_{01}, S), (P_i, T)) < \alpha$.

(2) Choose n such that P_1 is well approximated by a length n coding. That is, so that $|P_1 - P_1'| \ll \epsilon$, where P_1' is coded from P with a code of length n.

(3) Choose δ and m such that if (P', T') is any process and the m-distributions and entropies of (P', T') and (P, T) differ by at most δ, then $\overline{d}\big((P', T'), (P, T)\big) \ll \frac{\alpha}{n}$.

(4) Now let P_{01} be coded from Q by the same code with which P_1' is coded from P. By Exercise 640, the variation distance between Q and P_{01} is at most $|P - P_1'| + (2n + 1)\overline{d}((Q, S), (P, T)) < \frac{\epsilon}{2}$.

(5) By Exercise 639, $\overline{d}\big((P_{01}, S), (P_1', T)\big) \le (2n + 1)\overline{d}((Q, S), (P, T)) < \alpha$. □

6.6. Ornstein isomorphism theorem

Finally, we reap the harvest we have sown in the previous six chapters.

642. Theorem. *(Ornstein isomorphism theorem; Ornstein 1970a.) Any two finitely determined processes with the same entropy are isomorphic.*

Idea of proof. We are given transformations T and S and partitions P and Q such that (P, T) and (Q, S) are finitely determined of the same entropy, where

[102] We also require of α that for any partition Q' on the space of S, if $\overline{d}((P_1, T), (Q', S)) < \alpha$ then $H(S) - H(Q', S) \gg h(\alpha^{\frac{1}{2}})$. This will allow us to continue the proof of Sinai's theorem once we have chosen P_{01}; the reader who worked out the details is by now familiar with such considerations.

P generates. Our goal is to establish a Q' which also generates T such that (Q', T) has the same distribution as (Q, S).

(1) By Sinai, get q such that (q, T) has the same distribution as (Q, S).

    By Theorem 594, it suffices to get q' such that (q', T) has the same distribution as (Q, S), q' is arbitrarily close to q, and q' codes P arbitrarily well.

(2) Using Exercise 617, copy P in the land of q to get P', so that

    (i) $(P' \vee q, T)$ resembles $(P \vee q, T)$ in distribution and entropy. By the strong form of Sinai's theorem, by moving the copy a little after making it we can assume

    (ii) (P', T) has the same distribution as (P, T).

(3) Copy q to get q', so that

    (i) $(P \vee q', T)$ look like $(P' \vee q, T)$ in distribution and entropy. By the strong form of Sinai's theorem, by moving the copy a little after making it, we can assume

    (ii) (q', T) has the same distribution as (Q, S).

What all this means:

Since P generates, there is some n such that P codes q well by time n. $(P' \vee q, T)$, $(P \vee q, T)$ and $(P \vee q', T)$ all look alike so P codes q like P codes q' by time n. Thus q is close to q'.

P' is in the land of q so there is an n_1 such that q codes P' extremely well by time n_1. Hence, by making the copy q' good enough, we can guarantee that q' codes P extremely well by time n_1.

We are done. However there is a technicality to be concerned about when we play this game. It takes $2n_1 + 1$ letters from the q' partition to code a letter from P, when we are using an n_1 code. If any of those $2n_1 + 1$ letters are altered, the code codes wrong. Therefore, if we move the q' partition an amount on the order of $\frac{1}{n_1}$, that is enough to wreck the coding. Furthermore, you will recall that we did in fact move q' using the strong form of Sinai's theorem.

Relax. We did not make the q' coding until after we already knew n_1, and by making it arbitrarily well, we can get the distance we needed to move it to be small, even in comparison to $\frac{1}{n_1}$.

Here are some more details.

Sketch of proof. Let (P, T) and (Q, S) be finitely determined and of the same entropy, where P generates. Our goal is to find a partition Q' on the space of T which generates, such that $(Q', T) \approx (Q, S)$.

1. By Sinai's theorem, there is a partition q on the space of T such that $(q, T) \approx (Q, S)$.

643. Exercise. It suffices to show that for every $\epsilon > 0$ and $n \in \mathbf{N}$ there exists a partition q' on the space of T such that $(q', T) \approx (Q, S)$, $|q' - q| < \epsilon$, and q' codes P ϵ-well by time n. *Hint: Borel–Cantelli and Theorem 594.* •

2. Fix n and ϵ per the foregoing exercise and choose l so that P codes q $\frac{\epsilon}{10}$-well by time l. Pick a big $m \gg n$ and a small $\delta \ll \frac{\epsilon}{l}$. Using Exercise 617, copy P in the land of q to get P' so that $(P' \vee q, T)$ m-resembles $(P \vee q, T)$ up to δ and such that $H(P', T) > H(P, T) - \delta$. If you picked δ and m well enough, then the strong form of Sinai's theorem says you can move P' a little bit (way less than $\frac{\epsilon}{l}$, and still in the land of q!) to make $(P', T) \approx (P, T)$.

3. Choose $m' \gg m$ such that q codes P' $\frac{\epsilon}{10}$-well by time m'. Now choose $m'' \gg m'$ and $\delta' \ll \delta$; again using Exercise 617, copy q onto the space of T to get a partition q' such that $(P \vee q', T)$ m''-resembles $(P \vee q, T)$ up to δ' and $H(q', T) > H(q, T) - \delta'$. If you picked δ' and m'' well enough, then the strong form of Sinai's theorem says you can move q' a little bit (way less than $\frac{\epsilon}{m'}$) to make $(q', T) \approx (q, T)\big(\approx (Q, S)\big)$.

644. Exercise. Show that $|q - q'| < \epsilon$. *Hint: P codes q $\frac{\epsilon}{10}$-well by time l and $(P \vee q', T)$ $10l$-resembles $(P \vee q, T)$ up to $\frac{\epsilon}{10l}$.*

645. Exercise. Show that q' codes P ϵ-well by time n. *Hint: q codes P' $\frac{\epsilon}{10}$-well by time m' and $(P' \vee q', T)$ $10m'$-resembles $(P' \vee q, T)$ up to $\frac{\epsilon}{10m'}$.* □

646. Abbreviation. By Sinai (Q, S) isomorphic to factor (q, T) of (P, T). You would like to move q slightly to get an isomorphic copy (q', T) of (Q, S) which generates (P, T) better. Make (P', T) factor of (q, T): $(P' \vee q, T)$ looks like $(P \vee q, T)$. Make q' so that $(P \vee q', T)$ looks like $(P' \vee q, T)$. q' is close to q because (P, T) codes them about the same. (q', T) captures (P, T) well because it copies the way (q, T) captures (P', T).

647. Corollary. *Ornstein isomorphism theorem (classical form). Any two independent processes of the same entropy are isomorphic.*

648. Corollary. $FD \subset B$.

Proof. Let (P, T) be finitely determined. Choose an independent process (Q, S) of the same entropy. (Q, S) is finitely determined, therefore (P, T) and (Q, S) are isomorphic. □

7

Varieties of mixing

649. Discussion. In Chapters 5 and 6 we gave various randomness conditions on a stationary process X and showed that each was both necessary and sufficient for X to be isomorphic to a "maximally random" (i.e. independent) process. In this chapter, we will give a series of progressively weaker randomness conditions, showing each in turn to be necessary, but not sufficient, for the previous. As before, we use a boldfaced designation for the class of stationary processes satisfying a given condition, and formulate a "boldfaced theorem" specifying the various proper inclusions.

7.1. The varieties of mixing

In this subchapter we define the various classes. As always, the systems we consider are assumed to be invertible and Lebesgue.

650. Definition. We denote by **B** the class of Bernoulli systems and by **E** the class of ergodic systems.

651. Definition. A system $(\Omega, \mathcal{A}, \mu, T)$ is *Kolmogorov*, or K, if there is a sub-σ-algebra $\mathcal{K} \subset \mathcal{A}$ such that $T^{-1}\mathcal{K} \subset \mathcal{K}$, $\bigvee_{i=0}^{\infty} T^i \mathcal{K} = \mathcal{A}$, and $\mu(A) \in \{0, 1\}$ for all $A \in \bigcap_{i=0}^{\infty} T^{-i} \mathcal{K}$. We denote the class of Kolmogorov systems by **K**.

652. Definition. Let $E \subset \mathbf{N}$. The *upper density* of E is the number

$$\overline{d}(E) = \limsup_{n \to \infty} \frac{|E \cap \{1, 2, \ldots, N\}|}{N}.$$

If (x_n) is a sequence in a topological space, we write $D - \lim x_n = x$ if for every neighborhood U of x, there is a set E with $\overline{d}(E) = 0$ such that $x_n \in U$ for all $n \in (\mathbf{N} \setminus E)$.

653. Definition. Let (d_i) be a sequence of natural numbers. The set

$$R = \{d_{i_1} + d_{i_2} + \cdots + d_{i_k} : i_1 < i_2 < \cdots < i_k\}$$

is called an *IP set*. A set $E \subset \mathbf{N}$ is said to be IP* if it intersects every IP set non-trivially.

654. Definition. Let (x_n) be a sequence in a topological space. We write
IP* $\lim_n x_n = x$ if for every neighborhood U of x, the set $\{n : x_n \in U\}$ is
an IP* set.

655. Definition. Let $(\Omega, \mathcal{A}, \mu, T)$ be an invertible measure-preserving system. Then

(a) $(\Omega, \mathcal{A}, \mu, T)$ is *strongly mixing* if for any $A, B \in \mathcal{A}$, one has

$$\lim_{n \to \infty} \mu(A \cap T^n B) = \mu(A)\mu(B).$$

We denote the class of strongly mixing measure preserving systems
by **SM**.

(b) *(Furstenberg and Weiss* 1977.*)* $(\Omega, \mathcal{A}, \mu, T)$ is *mildly mixing* if for any
$A, B \in \mathcal{A}$, one has

$$\text{IP}^* \lim_{n \to \infty} \mu(A \cap T^n B) = \mu(A)\mu(B).$$

We denote the class of mildly mixing measure preserving systems by **MM**.

(c) $(\Omega, \mathcal{A}, \mu, T)$ is *weakly mixing* if for any $A, B \in \mathcal{A}$, one has

$$D\text{–}\lim_{n \to \infty} \mu(A \cap T^n B) = \mu(A)\mu(B).$$

We denote the class of weakly mixing measure preserving systems
by **WM**.

Definitions in place, we devote the remainder of the chapter to a proof of the
following.

656. Theorem. B $\subsetneq$ K $\subsetneq$ SM $\subsetneq$ MM $\subsetneq$ WM $\subsetneq$ E.

7.2. Ergodicity vs. weak mixing

In this brief subchapter, we show that the weakly mixing systems form a proper
subset of the ergodic systems, then give an alternate characterization of weak
mixing.

657. Theorem. WM $\subsetneq$ E.

Idea of proof. We use the following fact.

658. Exercise. Suppose that (a_n) is a bounded sequence of non-negative
numbers. If $D\text{-}\lim_n a_n = a$ exists, then $\lim_N \frac{1}{N} \sum_{n=1}^{N} a_n = a$. •

Suppose that $(\Omega, \mathcal{A}, \mu, T)$ is weakly mixing and let $A \in \mathcal{A}$ with $T^{-1}A = A$.
By the above exercise, $\mu(A)^2 = \lim_N \frac{1}{N} \sum_{n=1}^{N} \mu(A \cap T^{-n}A) = \mu(A)$. Hence
$\mu(A) \in \{0, 1\}$, which shows that $(\Omega, \mathcal{A}, \mu, T)$ is ergodic.

Now we give an example of an ergodic system that is not weakly mixing. The system is $(X, \mathcal{A}, \mu, T)$, where $(X, \mathcal{A}, \mu)$ is $[0, 1)$ with Lebesgue measure, and $Tx = x + \alpha \mod 1$, where α is irrational. The necessary properties will fall out of the following exercise.

659. Exercise. Show that, for this system, all forward orbits are dense. In fact, given ϵ, there is an M_ϵ such that all forward M_ϵ-orbits are ϵ-dense. *Hint: first reduce the problem to showing that the forward orbit of 0, namely $\{n\alpha : n \in \mathbf{N}\}$, comes ϵ-close to 0. Do this with the pigeonhole principle.* •

660. Exercise. Show that $(X, \mathcal{A}, \mu, T)$ is not weakly mixing. *Hint: let $A = [0, \frac{1}{3})$ and show that the set of n for which $(A \cap T^{-n}A) = \emptyset$ has density at least $\frac{1}{M_{\frac{1}{3}}}$.* •

661. Exercise. Show that $(X, \mathcal{A}, \mu, T)$ is ergodic. *Hint: let $A \in \mathcal{A}$ with $0 < \mu(A) < 1$. There exist intervals I and J almost contained in A and A^c, respectively. Use density of orbits to bring I to J.* ☐

There are many equivalent characterizations of weak mixing. We now prepare for a theorem that will establish equivalence of three of the most important of these.

662. Exercise. Prove that $(X, \mathcal{A}, \mu, T)$ is ergodic if and only if for all $f, g \in L^2(X)$, if one of $\int f \, d\mu$, $\int g \, d\mu$ is zero, then $D\text{-}\lim_n \int f T^n g \, d\mu = 0$. *Hint: approximate f and g by simple functions. For the converse, let $f = 1_A - \mu(A)$, $g = 1_B - \mu(B)$.*

663. Exercise. Let (a_n) be a sequence of real numbers. Show that $D\text{-}\lim_n a_n = 0$ if and only if $\lim_N \frac{1}{N} \sum_{n=1}^N a_n^2 = 0$.

664. Theorem. Let $(X, \mathcal{A}, \mu, T)$ be a measure-preserving system. The following are equivalent.

(a) $(X, \mathcal{A}, \mu, T)$ is weakly mixing.
(b) $(X \times X, \mathcal{A} \otimes \mathcal{A}, \mu \times \mu, T \times T)$ is ergodic.[103]
(c) T has no non-constant measurable eigenfunctions.

Idea of proof.

665. Exercise. Prove implication (b)→(a). *Hint: suppose one of $\int f \, d\mu$, $\int g \, d\mu$ is zero, let $a_n = \int f T^n g \, d\mu$ and use the previous two exercises; apply the mean ergodic theorem to the function $g \otimes g(x, y) = g(x)g(y)$ in the product space.* •

---

[103] Here $T \times T(x, y) = (Tx, Ty)$.

(a)→(c): First note the system is ergodic. Suppose there is a non-constant eigenfunction f; $Tf = \lambda f$. By ergodicity $\lambda \neq 1$, however $|\lambda| = 1$ since T is measure-preserving, hence $|f|$ is T-invariant and hence constant. However f is not constant, so

$$\int f\bar{f}\, d\mu = \int |f|^2\, d\mu = \left(\int |f|\, d\mu\right)^2 > \left|\int f\, d\mu\right|^2 = \left(\int f\, d\mu\right)\left(\int \bar{f}\, d\mu\right).$$

Hence there is an $\epsilon > 0$ and a neighborhood U of 1 such that for all $\alpha \in U$,

$$\left|\int \alpha f\bar{f}\, d\mu - \left(\int f\, d\mu\right)\left(\int \bar{f}\, d\mu\right)\right| > \epsilon.$$

One obtains a contradiction by the fact that for a set of n having positive density, $\lambda^n \in U$, a contradiction.

(c)→(b): We prove the contrapositive. Suppose (b) fails. Then there is a non-constant, $(T \times T)$-invariant function $H(x, y)$ in $L^2(X \times X, \mu \times \mu)$, which we may assume is bounded. For $f \in L^2(X)$, define $H * f$ by

$$H * f(x) = \int H(x, y) f(y)\, d\mu(y).$$

666. Exercise. Show that $f \to H * f$ is a bounded linear operator. *Hint: in fact the norm of this operator is at most $\|H\|_2$; use Fubini's theorem and Cauchy–Schwarz.* •

Indeed, much more is true: as an operator, H is in fact *compact*: that is, it takes bounded sets to precompact sets. Our next step is to construct a self-adjoint operator having the same properties. Assume that $H(x, y)$ is not a function of y alone (otherwise, H is not a function of x alone and proceed similarly). Put

$$K(x, y) = \int H(x, t)\overline{H(y, t)}\, d\mu(t).$$

By our assumption, K is not constant. Note that

$$K(Tx, Ty) = \int H(Tx, t)\overline{H(Ty, t)}\, d\mu(t)$$

$$= \int H(Tx, Tt)\overline{H(Ty, Tt)}\, d\mu(t)$$

$$= \int H(x, t)\overline{H(y, t)}\, d\mu(t) = K(x, y).$$

Moreover, as is obvious from the definition, K is self-adjoint: $K(y, x) = \overline{K(x, y)}$. Therefore, by the spectral theorem for compact self-adjoint operators, $L^2(X)$ has an orthonormal basis consisting of eigenvectors for K. Moreover, the eigenspace V_λ of each eigenvalue $\lambda \neq 0$ is of finite dimension.

667. Exercise. Show that, as operators, K and T commute. Conclude that T leaves each eigenspace V_λ invariant. ●

668. Exercise. Show that, restricted to the orthocomplement of the constants, K is not the zero operator. *Hint: write* $K_x(y) = K(x, y)$. *Use separability of* $L^2(X)$ *to find* f *not orthogonal to a.e.* K_x *and consider* $K * f$. ●

Consider now an eigenspace V_λ, where $\lambda \neq 0$, that is orthogonal to the constants. Since it is finite dimensional and T-invariant, it contains a one-dimensional T-invariant subspace, which must be spanned by a non-constant eigenfunction for T. □

7.3. Weak mixing vs. mild mixing

Next we show that the mildly mixing systems form a proper subset of the weakly mixing systems. First, we give an alternate characterization of mild mixing. This requires some preparation.

669. Definition. Let $(X, \mathcal{A}, \mu, T)$ be a measure-preserving system and let $f \in L^2(X)$. We say that f is *rigid* if there exists a sequence (n_i) of natural numbers such that $\| f - T^{n_i} f \| \to 0$.

670. Definition. We denote the set of finite, non-empty subsets of $\mathbf{N}$ by $\mathcal{F}$. If $\alpha, \beta \in \mathcal{F}$, we write $\alpha < \beta$ if $\max \alpha < \min \beta$.

671. Definition. (*Furstenberg and Katznelson* 1985.) Let (α_i) be a sequence in $\mathcal{F}$ with $\alpha_i < \alpha_{i+1}$, $i \in \mathbf{N}$. The set $\mathcal{F}^{(1)} = \left\{ \bigcup_{i \in \beta} \alpha_i : \beta \in \mathcal{F} \right\}$ is called an *IP-ring*. If $\mathcal{F}^{(1)}$ is an IP-ring, a function n from $\mathcal{F}^{(1)}$ into a commutative group $(G, +)$ will be called an *IP system* if $n(\alpha \cup \beta) = n(\alpha) + n(\beta)$ whenever $\alpha, \beta \in \mathcal{F}^{(1)}$ with $\alpha < \beta$.

672. Theorem. (*Hindman 1974*). *Let* $\mathcal{F}^{(1)}$ *be an IP-ring, let* $r \in \mathbf{N}$ *and suppose* $\mathcal{F}^{(1)} = \bigcup_{i=1}^{r} C_i$. *Then some* C_i *contains an IP ring* $\mathcal{F}^{(2)}$.

(Proof omitted.)

673. Definition. Let $\mathcal{F}^{(1)}$ be an IP-ring, let (X, ρ) be a metric space and suppose $f : \mathcal{F}^{(1)} \to X$ is a function. Let $x \in X$ and suppose that for every $\epsilon > 0$ there is some $\alpha_0 \in \mathcal{F}$ such that for every $\alpha \in \mathcal{F}^{(1)}$ with $\alpha > \alpha_0$, $\rho\big(x, f(\alpha)\big) < \epsilon$. Then we will write $\operatorname*{IP-lim}_{\alpha \in \mathcal{F}^{(1)}} f(\alpha) = x$.

674. Exercise. In the previous definition, if (X, ρ) is compact then there must exist an IP-ring $\mathcal{F}^{(2)} \subset \mathcal{F}^{(1)}$ such that $\operatorname*{IP-lim}_{\alpha \in \mathcal{F}^{(2)}} f(\alpha)$ exists. *Hint: use Hindman's theorem and a diagonal argument.*

675. Exercise. Let $\mathcal{H}$ be a separable Hilbert space, let $\mathcal{F}^{(1)}$ be an IP-ring and let U be an IP system from $\mathcal{F}^{(1)}$ into the space of unitary operators on $\mathcal{H}$. Then there exists an IP-ring $\mathcal{F}^{(2)} \subset \mathcal{F}^{(1)}$ such that $P = \underset{\alpha \in \mathcal{F}^{(2)}}{\text{IP-lim}} \ U(\alpha)$ exists in the weak operator topology. Moreover, P is an orthogonal projection. *Hint: for existence of the limit, use the previous exercise and the fact that the weak topology is compact metric when restricted to the unit ball of $\mathcal{H}$. To show P is an orthogonal projection, it suffices to show $P^2 = P$.*

676. Theorem. *(Furstenberg and Weiss 1977.) Let $(X, \mathcal{A}, \mu, T)$ be a measure-preserving system. The following are equivalent.*

(a) $(X, \mathcal{A}, \mu, T)$ is mildly mixing.
(b) There are no non-constant rigid functions in $L^2(\mu)$.

Proof. (b)→(a): Suppose there are no non-constant rigid functions. Let $\mathcal{H}$ be the orthocomplement of the constants in $L^2(\mu)$. T acts unitarily on $\mathcal{H}$. For (a), it is sufficient to show that for $f, g \in \mathcal{H}$, one has

$$\text{IP}^* \lim_n \int f T^n g \, d\mu = 0.$$

Let $n : \mathcal{F} \to \mathbf{N}$ be an arbitrary IP system. For some IP-ring $\mathcal{F}^{(1)}$, $P = \text{IP-}\lim_\alpha T^{n(\alpha)}$ exists in the weak operator topology. By the previous exercise, P is an orthogonal projection.

677. Exercise. Show that Pg is rigid for all $g \in \mathcal{H}$. Conclude that $P \equiv 0$. •

Hence

$$\underset{\alpha \in \mathcal{F}^{(1)}}{\text{IP-lim}} \ \int f T^{n(\alpha)} g \, d\mu = \int (Pf)(Pg) \, d\mu = 0.$$

(a)→(b): Suppose f is a non-constant rigid function. Choose (n_i) such that $\|T^{n_i} f - f\| < 2^{-i}$ and put $n(\alpha) = \sum_{i \in \alpha} n_i$, $\alpha \in \mathcal{F}$. Choose a set S such that $A = f^{-1}(S)$ has measure strictly between 0 and 1.

678. Exercise. Show that $\text{IP-}\lim_\alpha \mu\big(A \cap T^{n(\alpha)} A\big) = \mu(A)$ and use this to derive a contradiction to (a). □

679. Theorem. MM⊂WM.

680. Exercise. Prove the theorem. *Hint: show that any eigenfunction is a rigid function.* □

Our next task is to give an example of a weakly mixing system that is not mildly mixing. We use cutting and stacking. Let T_1 be a tower having one rung; the interval $[0, 1)$. Having constructed T_n, construct T_{n+1} as follows.

Break T_n into $n + 1$ equal width columns $C_1, \ldots, C_{n+1}$. Stack the columns vertically, starting with $C_1, \ldots$, etc. Put a single spacer between C_n and C_{n+1}. For example, T_2 looks like:

$$[\tfrac{1}{2}, \tfrac{2}{2})$$

$$[\tfrac{2}{2}, \tfrac{3}{2})$$

$$[\tfrac{0}{2}, \tfrac{1}{2})$$

T_3 looks like:

$$[\tfrac{5}{6}, \tfrac{6}{6})$$

$$[\tfrac{8}{6}, \tfrac{9}{6})$$

$$[\tfrac{2}{6}, \tfrac{3}{6})$$

$$[\tfrac{9}{6}, \tfrac{10}{6})$$

$$[\tfrac{4}{6}, \tfrac{5}{6})$$

$$[\tfrac{7}{6}, \tfrac{8}{6})$$

$$[\tfrac{1}{6}, \tfrac{2}{6})$$

$$[\tfrac{3}{6}, \tfrac{4}{6})$$

$$[\tfrac{6}{6}, \tfrac{7}{6})$$

$$[\tfrac{0}{6}, \tfrac{1}{6})$$

Normalize to a probability measure; denote the system this cutting and stacking construction converges to by (X, T).

681. Exercise. Show that (X, T) is ergodic. *Hint: see the proof below that Chacon's transformation is ergodic.*

682. Theorem. (X, T) *is weakly mixing.*

Sketch of proof. Suppose not. Then there is a non-constant eigenfunction f. Let $\lambda \neq 1$ be the corresponding eigenvalue. Denote by $\mathcal{A}_n$ the algebra generated by the rungs of T_n. Since $\bigvee_n \mathcal{A}_n$ generates, there is an n such that f is very nearly constant on at least some rungs of T_n.[104] But, since $f(Tx) = \lambda f(x)$ a.e., in fact f is nearly constant on *every rung* of T_n.

683. Exercise. Show that λ^h is very nearly 1, where h is the height of T_n.[105] *Hint: consider the first two stacked columns comprising T_{n+1}.*   •

---

[104] That is to say, $\mu(f^{-1}(S) \cap R) > 0.99\mu(R)$, where $\mathrm{diam}(S) \ll \mathrm{dist}(\lambda, 1)$ and R is the rung in question.

[105] In other words, $\mathrm{dist}(\lambda^h, 1) \ll \mathrm{dist}(\lambda, 1)$.

684. Exercise. Show that λ^{h+1} is very nearly 1. Conclude that λ is arbitrarily close to 1 and obtain a contradiction.[106] *Hint: consider the last two stacked columns comprising T_{n+1}, which are separated by a spacer.*   □

685. Theorem. *(X, T) is not mildly mixing.*

Idea of proof. *Let $f \in L^2(X)$ be arbitrary and let $\epsilon > 0$. We will find $h \in \mathbf{N}$ such that $\|T^h f - f\| < \epsilon$. Choose a very large n such that $\|f - E(f|\mathcal{A}_n)\| < \frac{\epsilon}{10}$. Let g be equal to $E(f|\mathcal{A}_n)$ on T_n and 0 on the rest of the space. If n is big enough, $\|f - g\| < \frac{\epsilon}{5}$.*

686. Exercise. Show that $\|g - T^h g\| \leq \frac{2\|g\|}{n}$, where h is the height of T_n. Assuming n large enough, this is less than $\frac{\epsilon}{2}$. *Hint: consider how T_{n+1} is constructed.*   •

Hence $\|f - T^h f\| \leq \|f - g\| + \|g - T^h g\| + \|T^h g - T^h f\| < \epsilon$.   □

7.4. Mild mixing vs. strong mixing

We begin with an example, due to Chacon, of a mildly mixing transformation that is not strongly mixing. We use cutting and stacking. Begin with the interval $[0, \frac{2}{3})$, which we label T_1. To get the nth tower T_n, cut the $(n-1)$st tower T_{n-1} into three equal columns. Stack two of them, add a spacer, then stack the third on top of that. For example, T_2 looks like this:

$[\frac{4}{9}, \frac{6}{9})$
$[\frac{6}{9}, \frac{8}{9})$
$[\frac{2}{9}, \frac{4}{9})$
$[\frac{0}{9}, \frac{2}{9})$

T_3 looks like this:

$[\frac{16}{27}, \frac{18}{27})$
$[\frac{22}{27}, \frac{24}{27})$
$[\frac{10}{27}, \frac{12}{27})$
$[\frac{4}{27}, \frac{6}{27})$
$[\frac{24}{27}, \frac{26}{27})$
$[\frac{14}{27}, \frac{16}{27})$

[106] That is, show that $\text{dist}(\lambda^{h+1}, 1) \ll \text{dist}(\lambda, 1)$, so that $\text{dist}(\lambda^{h+1}, \lambda^h) \ll \text{dist}(\lambda, 1)$.

$$[\tfrac{20}{27}, \tfrac{22}{27})$$

$$[\tfrac{8}{27}, \tfrac{10}{27})$$

$$[\tfrac{2}{27}, \tfrac{4}{27})$$

$$[\tfrac{12}{27}, \tfrac{14}{27})$$

$$[\tfrac{18}{27}, \tfrac{20}{27})$$

$$[\tfrac{6}{27}, \tfrac{8}{27})$$

$$[\tfrac{0}{27}, \tfrac{2}{27})$$

687. Definition. *Chacon's transformation* is the transformation that the above cutting and stacking procedure converges to.

688. Theorem. *Chacon's transformation is ergodic.*

Proof. *Suppose* $A = TA$ *with* $\mu(A) \in (0, 1)$. *Let* $1 > \rho > \mu(A)$. *Using the existence of Lebesgue points of density, there is an interval* $I = [\tfrac{i}{3^n}, \tfrac{i+2}{3^n})$ *such that* $\mu(A \cap I) > |I|\rho$. *Notice that* I *is a rung of* T_n. *Since* $A = TA$, *every rung* J *of* T_n *satisfies* $\mu(A \cap J) > |J|\rho$. *Summing over all rungs, we get* $\mu(A) > \rho \tfrac{3^n - 1}{3^n}$. *Choosing* n *large enough, we obtain a contradiction.* $\square$

689. Theorem. *Chacon's transformation is mildly mixing.*

Idea of proof. Suppose there is a non-constant rigid function f. We shall obtain a contradiction.

690. Exercise. Show there is a proper measurable A such that 1_A is rigid. *Hint: take* $A = f^{-1}(S)$ *for some measurable S.* •

By ergodicity, $\mu(A \triangle TA) = \rho > 0$. Let $\mathcal{A}_n$ denote the σ-algebra generated by the rungs of T_n. Then $\bigvee_n \mathcal{A}_n$ generates, hence there is an m and a set B consisting of various rungs of T_m such that $\mu(A \triangle B) < \tfrac{\rho}{10^6}$.

Let h_n denote the height of T_n. Choose $n \gg h_m$ such that $\mu(A \triangle T^n A) < \tfrac{\rho}{10^6}$. Then $\mu(B \triangle T^n B) < \tfrac{\rho}{10^5}$. Choose t with $10n < h_t < 100n$. Note that B is the union of some rungs of $\mathcal{A}_t$. Let $b_1, \ldots, b_{h_t}$ be in $\{0, 1\}$ such that $b_i = 1$ if and only if the ith rung of A_t is in B. Let s_1 be the number of i, $1 \le i \le n$, such that $b_i \ne b_{h_t - n + i}$.

691. Exercise. Show that $s_1 < \tfrac{\rho n}{10^3}$. *Hint: observe how tower* A_{t+1} *is built, in particular how the middle third of* A_t *is stacked atop the left third; use the fact that* $\mu(B \triangle T^n B)$ *is small and* $n > \tfrac{h_t}{100}$. •

Next, let s_2 be the number of i, $1 \le i \le n$, such that $b_i \ne b_{h_t - n + i + 1}$.

692. Exercise. Show that $s_2 < \frac{\rho n}{10^3}$. *Hint: again observe how tower A_{t+1} is built, this time paying attention to how the right third of A_t is separated from the middle third by a single spacer.* •

Note $h_{t-5} < n$. Let s_3 be the number of i, $1 \le i \le h_{t-5}$, such that $b_i \ne b_{i+1}$.

693. Exercise. Show that $s_3 \le s_1 + s_2 < \frac{\rho n}{500} < \frac{\rho h_{t-5}}{10}$. *Conclude that* $\mu(B \triangle TB) < \frac{\rho}{5}$ *and derive a contradiction. Hint: B is a union of rungs of T_{n-5}, too.* □

694. Theorem. *Chacon's transformation is not strongly mixing.*

Idea of proof. Let $A = [0, \frac{2}{9})$, i.e. the base of T_2.

695. Exercise. Show that for all n, $\mu(A \cap T^{h_n} A) \ge \frac{2}{9} \cdot \frac{1}{3}$. *Hint: for $x \in A$, if x lies in the leftmost third of T_n then $T^{h_n} x$ will lie in the middle third of the same rung of T_n.* □

7.5. Strong mixing vs. Kolmogorov

696. Theorem. $K \subset SM$.

Idea of proof. *Let $(X, \mathcal{A}, \mu, T)$ be a Kolmogorov system; suppose it is not mixing. Then there are measurable sets A, B and a sequence n_i such that $\mu(A \cap T^{n_i} B) \to x \ne \mu(A)\mu(B)$. There is a sub-$\sigma$-algebra $\mathcal{K} \subset \mathcal{A}$ such that $T^{-1}\mathcal{K} \subset \mathcal{K}$, $\bigvee_{i=0}^{\infty} T^i \mathcal{K} = \mathcal{A}$, and $\mu(A) \in \{0, 1\}$ for all $A \in \bigcap_{i=0}^{\infty} T^{-n}\mathcal{K}$. Since $\bigvee_{i=0}^{\infty} T^i \mathcal{K} = \mathcal{A}$, we can approximate B very closely by a set C in $\bigvee_{i=0}^{n} T^i \mathcal{K} = T^n \mathcal{K}$ for some n and still have $\mu(A \cap T^{n_i} C) \to y \ne \mu(A)\mu(C)$. Passing to a subsequence if necessary, we can assume $1_{T^{-n_i} C} = T^{n_i} 1_C \to f$ in the weak topology.*[107]

697. Exercise. Use the fact that $T^{-1}\mathcal{K} \subset \mathcal{K}$ to show that f is measurable with respect to $T^{n-n_i}\mathcal{K}$ for every i; use the fact that $\bigcap_{i=0}^{\infty} T^{-n}\mathcal{K}$ is trivial to conclude that f is constant and obtain a contradiction. *Hint: first show that $\int f = \mu(C)$.* □

In the remainder of this subchapter, we demonstrate that there are strongly mixing systems that do not have the Kolmogorov property. This will be done in three steps. First, we show that any K system has positive entropy. Next, we show that any rank-one system has zero entropy. Finally, we exhibit a mixing rank-one system.

---

[107] That is, we may assume that $\langle T^{n_i} 1_C, g \rangle \to \langle f, g \rangle$ for every $g \in L^2(X)$. This is easy to show elementarily by considering a dense subset of $L^2(X)$ and using a diagonal argument.

698. Theorem. *Every Kolmogorov system has positive entropy.*

Idea of proof. Let $(X, \mathcal{A}, \mu, T)$ be a Kolmogorov system, and suppose this system has zero entropy. There is a sub-σ-algebra $\mathcal{K} \subset \mathcal{A}$ such that $T^{-1}\mathcal{K} \subset \mathcal{K}$, $\bigvee_{i=0}^{\infty} T^i \mathcal{K} = \mathcal{A}$, and $\mu(A) \in \{0, 1\}$ for all $A \in \bigcap_{i=0}^{\infty} T^{-n}\mathcal{K}$. Let $A \in \mathcal{A}$ be arbitrary. Since $\bigvee_{i=0}^{\infty} T^i \mathcal{K} = \mathcal{A}$, we can approximate A very closely by a set C in $\bigvee_{i=0}^{n} T^i \mathcal{K} = T^n \mathcal{K}$ for some n. Let $P = \{C, X \setminus C\}$.

699. Exercise. Show that $\bigvee_{i=-\infty}^{m} T^i P \subset \bigvee_{i=-\infty}^{m-1} T^i P$ mod 0 for all $m \in \mathbf{Z}$. *Hint: by zero entropy, the past determines the present for the (P, T) process.*

                                                       •

Iterating the previous exercise, we get $C \in \bigcap_{i=0}^{\infty} T^{-n}\mathcal{K}$. But this σ-algebra is trivial.   □

700. Exercise. Let $n \in \mathbf{N}$ be large and let w be a word of length n on the alphabet $\{0, 1\}$. Then the number of words v of length m having the property that a proportion close to 1 of v can be exhausted by disjoint copies of w is exponentially small. [108]

701. Theorem. *Rank-one systems have zero entropy.*

Idea of proof. Let $(X, \mathcal{A}, \mu, T)$ be a rank-one system constructed from a sequence of towers T_n of heights h_n.

702. Exercise. It suffices to show that the (P, T) process has zero entropy for every 2-set partition P.   •

Let $A \in \mathcal{A}$ be arbitrary and put $P = \{A, X \setminus A\}$. Suppose that $H(P, T) > 0$. Choose n such that A is approximable by a set B that is a union of rungs of T_n. Setting $Q = \{B, X \setminus B\}$, we will have $H(Q, T) > 0$ (provided the approximation is good enough). Without changing the set B, we can assume that n is as large as desired (that is, B will still be a union of rungs of T_n). In particular, we can assume that most of the space is taken up by T_n. Let w be the Q-name of length n of the base of T_n.

703. Exercise. Show that for $m \gg h_n$, for most points x, the (Q, m)-name of x contains nearly $\frac{m}{h_n}$ disjoint copies of w. *Hint: consider a tower T_k with $h_k \gg m$; most rungs of the tower have m-names consisting mostly of passes through T_n.*   •

---

[108] A more precise formulation: let $\epsilon > 0$ and let $m \gg n$. The number of words of length m having at least $\frac{m(1-\epsilon)}{n}$ disjoint copies of w as subwords is at most $3^{(\frac{m}{n}+m\epsilon)}$. *Hint: consider the map taking words on $\{0, 1, 2\}$ to words on $\{0, 1\}$ induced by sending 2 to a copy of w.*

By the previous exercise, Exercise 700 and the fact that n can be taken arbitrarily large, the number of (Q, m)-names is as exponentially small as desired, which contradicts $H(Q, T) > 0$. □

We now give an example of a mixing rank-one transformation. The result is due to T. Adams, based on a conjecture of Smorodinsky. We need the following version of the ergodic theorem (we give two formulations, with two suggested proofs).

704. Exercise. Let $(X, \mathcal{A}, \mu, T)$ be ergodic. Show that

$$\left| \frac{1}{n} \sum_{i=1}^{n} \mu(T^{-i} A \cap B) - \mu(A)\mu(B) \right| \to 0$$

uniformly over sets A as $n \to \infty$. *Hint: first show that $\frac{1}{n} \sum_{i=1}^{n} \int f T^{-i} g \, d\mu \to 0$ uniformly over functions f of bounded norm by taking the sum inside and employing Cauchy–Schwarz. Take $f = 1_A$ and $g = 1_B - \mu(B)$.*

705. Exercise. Let $(X, \mathcal{A}, \mu, T)$ be an ergodic measure-preserving system, let $B \in \mathcal{A}$ and let $(A_n) \subset \mathcal{A}$. Then $\left| \lim_{n \to \infty} \frac{1}{n} \sum_{i=0}^{n-1} \mu(T^{-i} A_n \cap B) - \mu(A_n)\mu(B) \right| = 0$. *Hint: use convergence in measure in the ergodic theorem. For nearly all x, approximately $\mu(B)n$ members of the set $\{x, Tx, \ldots, T^{n-1}x\}$ lie in B; in particular, for nearly all $x \in A_n$.*

706. Definition. Define a system by cutting and stacking as follows. Let $T_1 = [0, 1)$. We view this as a tower having a single rung. To get T_2 from T_1, cut T_1 into two subcolumns and stack them, putting a spacer in between. T_2 looks like this:

$$[\tfrac{1}{2}, \tfrac{2}{2})$$

$$[\tfrac{2}{2}, \tfrac{3}{2})$$

$$[\tfrac{0}{2}, \tfrac{1}{2})$$

To get T_3 from T_2, cut T_2 into three subcolumns and stack them, putting one spacer in between the first two subcolumns and two spacers between the last two subcolumns. T_3 looks like this:

$$[\tfrac{5}{6}, \tfrac{6}{6})$$

$$[\tfrac{8}{6}, \tfrac{9}{6})$$

$$[\tfrac{2}{6}, \tfrac{3}{6})$$

$$[\tfrac{11}{6}, \tfrac{12}{6})$$

$$[\tfrac{10}{6}, \tfrac{11}{6})$$

$$[\tfrac{4}{6}, \tfrac{5}{6})$$

$$[\tfrac{7}{6}, \tfrac{8}{6})$$

$$[\tfrac{1}{6}, \tfrac{2}{6})$$

$$[\tfrac{9}{6}, \tfrac{10}{6})$$

$$[\tfrac{3}{6}, \tfrac{4}{6})$$

$$[\tfrac{6}{6}, \tfrac{7}{6})$$

$$[\tfrac{0}{6}, \tfrac{1}{6})$$

Having constructed T_{n-1}, cut it into n subcolumns $C_1, \ldots, C_n$ and stack them, putting i spacers between C_i and C_{i+1}. This gives T_n; denote its height by h_n. Finally, normalize the measure; the resulting system $(X, \mathcal{A}, \mu, T)$ is called *the staircase*.

707. Exercise. Show that the staircase is totally ergodic.[109] *Hint: suppose $T^j B = B$. Choose $n \gg j$ and approximate B by a measurable set A that is a union of rungs of T_n. For $kj \approx h_n$, a typical rung I of T_n will be such that $T^{kj} I$ is evenly dispersed among about n rungs of T_n. Show that $0 < \mu(B) < 1$ is inconsistent with $T^{kj} A \approx A$.*

708. Exercise. Let $(X, \mathcal{A}, \mu, T)$ be a measure-preserving system, let $B \in \mathcal{A}$ and let $R, L, \rho \in \mathbf{N}$. Then

$$\int \left| \frac{1}{R} \sum_{i=0}^{R-1} 1_B(T^{-i} x) - \mu(B) \right| d\mu(x) \leq \int \left| \frac{1}{L} \sum_{i=0}^{L-1} 1_B(T^{-i\rho} x) - \mu(B) \right| d\mu(x) + \frac{\rho L}{R}.$$

Hint: obvious if $\frac{\rho L}{R} > 1$. Otherwise, tile most of $\{1, \ldots, R\}$ by progressions $\{a, a + \rho, \ldots, a + \rho(L-1)\}$.

709. Exercise. Let $\epsilon > 0$. There exists $\delta > 0$ such that if $(X, \mathcal{A}, \mu, T)$ is a probability space and $f \in L^2(X)$ then $\int |f|^2 \, d\mu < \delta$ implies $\int |f| \, d\mu < \epsilon$.

710. Theorem. *Suppose $(X, \mathcal{A}, \mu, T)$ is the staircase. Let B be a union of levels in some column and suppose $l_n, \rho_n \to \infty$ with $h_n \leq \rho_n \leq 2h_n$ (where h_n is the height of the nth column). Then*

$$\lim_{n \to \infty} \int \left| \frac{1}{l_n} \sum_{i=0}^{l_n-1} 1_B(T^{-i\rho_n} x) - \mu(B) \right| d\mu(x) = 0.$$

---

[109] That is, T^j is ergodic for all j.

Proof. Let $\epsilon > 0$. Choose $\delta > 0$ as in the foregoing exercise. Fix i and write $i\rho_n = jh_n + t$, where $0 \leq t < h_n$. If n is sufficiently large then B is a union of levels of T_n. Write $B \approx B_1 \cup B_2$, where B_1 consists of that part of B lying in the top $t - (j+1)(n+1)$ levels of T_n minus the rightmost $(j+1)$ subcolumns of T_n, and B_2 consists of that part of B lying in the rest of T_n, minus the rightmost $(j+1)$ subcolumns and bottom $j(n+1)$ levels of T_n.

Let I be that part of a level that lies in B_1.

711. Exercise. Show that I passes through the staircase $(j+1)$ times under $T^{i\rho_n}$. Conclude that $T^{i\rho_n} I$ intersects $(n-j)$ levels of T_n, lying in an arithmetic progression with gap $j+1$, with measure $\frac{\mu(I)}{n+1}$ each. •

Denote the topmost level that $T^{i\rho_n} I$ intersects with measure $\frac{\mu(I)}{n+1}$ by I^*, and put $B_1^* = \bigcup_{I \subset B_1} I^*$. Then

$$\mu(T^{i\rho_n} B_1 \cap B) = \frac{1}{n+1} \sum_{k=0}^{n-j-1} \mu(T^{-k(j+1)} B_1^* \cap B).$$

712. Exercise. Carry out a similar construction for B_2^* such that

$$\mu(T^{i\rho_n} B_2 \cap B) = \frac{1}{n+1} \sum_{k=0}^{n-j-1} \mu(T^{-kj} B_2^* \cap B).$$

•

713. Exercise. Conclude that $\mu(T^{i\rho_n} B \cap B) \to \mu(B)^2$ as $n \to \infty$. *Hint: the B_k^* depend on n, however T^j and T^{j+1} are ergodic and Exercise 705 applies.*

•

By the foregoing exercise, $\mu(T^{i\rho_n} B \cap T^{k\rho_n} B) \to \mu(B)^2$ as $n \to \infty$ for $i \neq k$.

714. Exercise. Use this fact to show that for all sufficiently large L, one has for large n,

$$\int \left| \frac{1}{L} \sum_{i=0}^{L-1} 1_B(T^{-i\rho_n} x) - \mu(B) \right|^2 d\mu < \delta.$$

Hint: just expand the product and take limits. •

By the choice of δ, one has

$$\limsup_{n \to \infty} \int \left| \frac{1}{l_n} \sum_{i=0}^{l_n - 1} 1_B(T^{-i\rho_n} x) - \mu(B) \right| d\mu \leq \epsilon.$$

Since ϵ is arbitrary, we are done. □

715. Theorem. *(Adams 1998.) The staircase is mixing.*

Idea of proof. Recall that h_n is the height of T_n. Note that $h_n > n!$.

716. Exercise. Show that it suffices to show that $\mu(T^{m_n} A \cap B) \to \mu(A)\mu(B)$ for all sets A, B that are a union of rungs of some T_k and for an arbitrary sequence (m_n) satisfying $h_n \le m_n < h_{n+1}$. •

Let A, B, (m_n) be as in the preceding exercise. Pick n large. Label the subcolumns of T_n $C_1, \ldots, C_{n+1}$. Write $m = m_n$ and let $m = kh_n + t$, where $1 \le k \le n$ and $0 \le t \le h_n$. Our plan is to break T_n into three pieces, D_1, D_2 and D_3, let A_i be (more or less) the parts of A lying in D_i, and show that $\mu(T^{m_n} A_i \cap B) \to \mu(A_i)\mu(B)$, $i = 1, 2, 3$.

Let D_1 consist of $C_{n-k}, C_{n-k+1}, \ldots, C_{n+1}$; that is, the rightmost $k + 1$ subcolumns of T_n. Recall that T_{n+1} is formed by stacking these subcolumns one atop the other, with t spacers between C_t and C_{t+1}, and that T acts by moving up T_{n+1}.

717. Exercise. Show that, under T^m, most rungs of T_{n+1} that lie in D_1 move through the top part of T_{n+1} exactly once, then through a variable number of spacers according to which subcolumn of T_{n+1} they are in, then into the bottom part of T_{n+1}. •

718. Exercise. Show that most rungs R of T_{n+1} that lie in D_1 have the property that $T^m R$ intersects $n + 2$ consecutive rungs of T_{n+1} in sets having measure $\frac{\mu(R)}{n+2}$ each. •

Call the rungs of T_{n+1} that lie in D_1 and satisfy the foregoing exercise "good rungs". Let A_1 be the part of A that consists of good rungs in D_1 and let A_1^* be the set of rungs of T_{n+1} consisting of the topmost rung intersecting $T^m R$ in a set having measure $\frac{\mu(R)}{n+2}$, for each good rung R in A_1. Of course one has $\mu(A_1^*) = \mu(A)$.

719. Exercise. By summing over all the rungs R in A_1, verify the following:

$$\mu(T^m A_1 \cap B) = \frac{1}{n+2} \sum_{i=0}^{n+1} \mu(T^{-i} A_1^* \cap B) \to \mu(A_1^*)\mu(B) = \mu(A_1)\mu(B).$$

Hint: use Exercise 704. •

Let now A_2 be that portion of A lying in all but the bottommost n^2 levels (in T_n) of D_2. (Note that when n is large, the union of these levels has very small measure.) We will show that $\lim_n \mu(T^{m_n} A_2 \cap B) = \mu(A_2)\mu(B)$ (recall that A_2 depends on n, though we are suppressing this in the notation). Assume without loss of generality that $\frac{k_n}{n} < 1 - \epsilon$ for some $\epsilon > 0$.

Let $I = A_2 \cap r$ for some rung r.

720. Exercise. Under T^{m_n}, I passes through the staircase topping T_n exactly k_n+1 times; hence $T^{m_n}I$ intersects $n-k_n-1$ levels of T_n with measure $\frac{\mu(I)}{n-k_n-1}$. Moreover, these levels lie in an arithmetic progression having gap size $k_n + 1$.

          •

Denote by I^* the (entire) topmost level of T_n that $T^{m_n}I$ intersects, and put $A^* = \bigcup_{I \subset A_2} I^*$. Notice that $\mu(A^*) = \frac{n}{n-k_n-1}\mu(A_2)$.

721. Exercise. Verify that $\mu(T^{m_n} A_2 \cap B) = \frac{1}{n} \sum_{i=0}^{n-k_n-2} \mu(T^{-i(k_n+1)} A^* \cap B)$.

          •

722. Exercise. In order to show that $\lim_n \mu(T^{m_n} A_2 \cap B) = \mu(A_2)\mu(B)$, it suffices to show that

$$\int \left| \frac{1}{n - k_n - 1} \sum_{i=0}^{n-k_n-2} 1_B(T^{-i(k_n+1)}x) - \mu(B) \right| d\mu(x) \to 0. \qquad (*)$$

Hint: integrate just over A, move the absolute value bars outside the integral and use the previous exercise.

          •

Choose $p = p_n$ such that $h_{p-1} \leq k_n + 1 \leq h_p$.

723. Exercise. Show that $\frac{(n-k_n-2)(k_n+1)}{h_p} \to \infty$. *Hint: recall we assumed that* $\frac{k_n}{n} < 1 - \epsilon$.

          •

Denote by r_n the smallest integer i such that $i(k_n + 1) \geq h_p$.

724. Exercise. There exists a sequence $l_n \to \infty$ such that $\frac{n-k_n-2}{l_n r_n} \to \infty$. *Hint: show first that* $\frac{n-k_n-2}{r_n} \to \infty$.

          •

By Theorem 710, one has

$$\lim_{n \to \infty} \int \left| \frac{1}{l_n} \sum_{i=0}^{l_n-1} 1_B(T^{-i r_n(k_n+1)}x) - \mu(B) \right| d\mu(x) = 0.$$

Applying now Exercise 708, one gets $(*)$.

    Finally, let A_3 be that part of A lying in D_3, excepting that part lying in the bottom n^2 rungs of T_n.

725. Exercise. Show that $\lim_n \mu(T^{m_n} A_3 \cap B) = \mu(A_3)\mu(B)$. *Hint: show that each rung of A_3 moves through the staircase k_n times under T^{m_n} and proceed as in the previous part.*

          •

    Since $T_n = D_1 \cup D_2 \cup D_3$, $A_i \approx A \cap D_i$ and $\lim_n \mu(T^{m_n} A_i \cap B) = \mu(A_i)\mu(B)$, $i = 1, 2, 3$, we can conclude that $\lim_n \mu(T^{m_n} A \cap B) = \mu(A)\mu(B)$.

          □

7.6. Kolmogorov vs. Bernoulli

Our first task in this section is to show that Bernoulli processes are
K-automorphisms.

726. Exercise. Show that the K property is an isomorphism invariant.

727. Theorem. $\mathbf{B} \subset \mathbf{K}$.

Proof. By the foregoing exercise, it suffices to establish that independent processes are K. Let (X_i) be an independent process on a finite alphabet Λ. Let
P be the partition determined by the zeroth coordinate, that is, into pieces
$X_0^{-1}(\lambda)$, $\lambda \in \Lambda$, and put $\mathcal{K} = \bigvee_{i=0}^{\infty} T^{-i} P$.

728. Exercise. Show that $T^{-1}\mathcal{K} \subset \mathcal{K}$, $\bigvee_{i=0}^{\infty} T^i \mathcal{K}$ generates $\mathcal{A}$, and $\mu(A) \in$
$\{0, 1\}$ for all $A \in \bigcap_{i=0}^{\infty} T^{-i}\mathcal{K}$. □

Our next task is to produce a system that is K but not Bernoulli. This
was originally done by Ornstein (1973), but we will give a more natural
example. The system is called the T, T^{-1} process, and is defined as follows.
Let $(Y_i)_{i=-\infty}^{\infty}$ be i.i.d. random variables taking values -1 and 1 each with
probability $\frac{1}{2}$. Next, let $(Z_i)_{i=-\infty}^{\infty}$ be i.i.d. random variables taking values
H and T each with probability $\frac{1}{2}$. Now put $X_0 = (Y_0, Z_0)$. For $n > 0$,
put $X_n = (Y_n, Z_s)$, where $s = \sum_{i=0}^{n-1} Y_i$, and put $X_{-n} = (Y_{-n}, Z_r)$, where
$r = \sum_{i=1}^{n} -Y_{-i}$.

729. Definition. $(X_i)_{i=-\infty}^{\infty}$, as defined in the previous paragraph, is the
T, T^{-1} *process.*

The reader should make note of how to generate an output of the T, T^{-1}
process. First, generate a random sequence (y_i) of 1s and -1s. This is called
the *random walk.* Next, generate a random sequence (z_i) of Ts and Hs. This
is called the *random scenery.* Start at the origin of both sequences and output
(y_0, z_0). Now march to the right in the sequence (y_i) (this generates the first
coordinate of the output), and at the same time do the random walk that (y_i)
generates along the scenery sequence (z_i) (this generates the second coordinate
of the output). Thus the T, T^{-1} process is often called a *random walk on a
random scenery.*

We now move to showing that the T, T^{-1} process is a K system. The
following theorem will provide a sufficient condition.

730. Theorem. *Let* $(X, \mathcal{A}, \mu, T)$ *be a measure-preserving system, and let* Q
be a generator. If for every measurable set A *one has* $\lim_{n \to \infty} P(T^n x \in$
$A | \bigvee_{j=-\infty}^{0} T^j Q) = P(A)$ *a.e., then* $(X, \mathcal{A}, \mu, T)$ *is a K system.*

Sketch of proof. Let $\mathcal{K} = \bigvee_{j=-\infty}^{0} T^j Q$. Then trivially $T^{-1}\mathcal{K} \subset \mathcal{K}$ and
$\bigvee_{i=0}^{\infty} T^i \mathcal{K} = \mathcal{A}$. Let $A \in \bigcap_{i=0}^{\infty} T^{-i}\mathcal{K}$. Then for arbitrary $n > 0$, one has

$A \in T^{-n}\mathcal{K} = \bigvee_{j=-\infty}^{-n} T^j Q$. It follows that $P\left(x \in A \mid \bigvee_{j=-\infty}^{-n} T^j Q\right)$ is 0, 1-valued. This implies that $P\left(T^n x \in A \mid \bigvee_{j=-\infty}^{0} T^j Q\right)$ is 0, 1-valued. Letting $n \to \infty$ and applying the hypothesis, we get that $P(A) \in \{0, 1\}$. □

731. Theorem. *Let $(X_i)_{i=-\infty}^{\infty}$ be the T, T^{-1} process. For any measurable set A, one has $\lim_{n\to\infty} P\left(X_n \in A \mid \bigvee_{i=-\infty}^{0} X_i\right) = P(A)$ a.e.*

Idea of proof. Suppose not. Then there is a set A and an $\epsilon > 0$ such that

$$\left| \lim_{n\to\infty} P\left(X_n \in A \mid \bigvee_{i=-\infty}^{0} X_i\right) - P(A) \right| > \epsilon$$

on a set of measure at least ϵ.

732. Exercise. Show that by approximation, we can assume without loss of generality that A is a cylinder set. •

Since the values of X_i, $i \le 0$, can obviously provide no information whatsoever about the first coordinate of X_M, we can in fact assume that A is a cylinder set that comes about by the specifying of second coordinates only. Indeed, here we will assume for convenience that $A = \{X_0 \in \{(-1, T), (1, T)\}\}$; readers should convince themselves that the argument to follow can be adapted to the general case.

733. Exercise. Let $\epsilon > 0$. There is an N such that for any sufficiently large M, the following occurs: with probability at least $1 - \epsilon$, a randomly chosen scenery (s_i) will have the property that one can cover all but ϵ (in relative counting measure) of $\{-M, -M+1, \ldots, M\}$ by intervals I of length N, such that in every such interval I, the proportion of $i \in I$ with $s_i = T$ is within ϵ of $\frac{1}{2}$. •

734. Exercise. Use the foregoing exercise to show that, given $\epsilon > 0$, for all sufficiently large M one has

$$\left| P\left(X_M \in \{(-1, T), (1, T)\} \mid \bigvee_{i=-\infty}^{0} X_i\right) - \frac{1}{2} \right| < \epsilon$$

on a set of measure at least $1 - \epsilon$. *Hint: let $M \gg N$ in the previous exercise, so that (most of) the binomial distribution $f(2x - M) = \binom{M}{x} 2^{-M}$ is essentially flat on intervals of length N.* □

The following is now immediate, hence we omit the proof.

735. Corollary. *The T, T^{-1} process is K.*

All that is left to do now is to show that the T, T^{-1} process is not Bernoulli. Unfortunately, the proof of this is rather involved, and, unlike some of the other involved arguments we did give in some detail, not particularly central in its methodology to ergodic theory as a whole. Accordingly, we will be far more brief and incomplete here than elsewhere in the book. The reader who wants more detail can consult Kalikow (1982).

736. Theorem. *The T, T^{-1} process is not Bernoulli.*

737. Vague idea of proof. Our plan is to show that the T, T^{-1} process doesn't satisfy the very weak Bernoulli condition. So, suppose it did. As a consequence of that, you and I could play the following game. First, we each choose (independently) a random scenery. Next, I choose a random path. Together with my scenery, this path generates an output. Now (assuming that the T, T^{-1} process satisfies the very weak Bernoulli condition), with very high probability, by changing less than *one one-billionth of the entries in my output*, it will be possible for me to convince you that my output comes from some path paired with *your scenery*! Think about that for a moment. In running through an output of the T, T^{-1} process, one becomes very, very intimately familiar with one's scenery. The random walk traipses over it again, and again, and again. There just *isn't any chance* that two completely independently chosen sceneries will look anywhere nearly enough alike that one could mistake an output using the one for an output using the other, even having changed a smallish fraction of the entries.

Idea of proof. Well, okay. So much for the vague idea. Unfortunately, as obviously sound as it seems, it turns out to be pretty tricky to make a rigorous proof out of the idea. We start with the following.

738. Exercise. Suppose (C_n) is a sequence of positive numbers with $\sum_{n=0}^{\infty} \frac{\log C_n}{2^{n+1}} < \infty$, for some $M > 0$. Let (A_n) be a sequence satisfying the recursion relation $A_{n+1} = C_n A_n^2$. There exists $\delta > 0$ such that if $A_0 < \delta$ then $A_n \to 0$ as $n \to \infty$. *Hint: after taking logarithms, divide through by 2^{n+1} and observe that $\frac{\log A_n}{2^n}$ telescopes. Let $\delta = \exp(-\sum_{n=0}^{\infty} \frac{\log C_n}{2^{n+1}})$.* •

Letting s_1, s_2 be two sceneries and p_1, p_2 be two paths, for $I = [a, b)$ an interval write $d_I\big((s_1, p_1), (s_2, p_2)\big)$ for the mean Hamming distance between the outputs generated by (s_1, p_1) and (s_2, p_2) restricted to the interval I. Then, as explained in the vague idea section, for the process to be very weak Bernoulli it would have to be the case that for fixed $\epsilon > 0$, and randomly chosen s_1, s_2 and p_2, there is usually a choice of p_1 such that $d_{[0,n)}\big((s_1, p_1), (s_2, p_2)\big) < \epsilon$ for sufficiently large n.

We will violate the above by constructing a decreasing sequence (ϵ_k) converging to some $\epsilon > 0$, an increasing sequence (n_k) of natural numbers, and sets R_k of "reasonable" paths of length n_k, which asymptotically contain a fraction of all paths tending to 1, such that letting

$$A_k = \max_{p_2 \in R_k, s_1} P\Big(\exists p_1 \text{ such that } d_{[0,n_k)}\big((s_1, p_1), (s_2, p_2)\big) < \epsilon_k\Big),$$

one has $A_k \to 0$ as $k \to \infty$. (The probability referred to is determined by the distribution of s_2, which is treated as being chosen randomly while s_1 and p_2 are fixed.)

Here are the definitions: let $n_k = 10^{18(k+1)^2}$ and choose $\epsilon = \frac{1}{3n_0}$. For $k \in \mathbf{N} \cup \{0\}$, put $\epsilon_k = \epsilon(1 + \frac{1}{k+1})$. Our next task is to define the sets R_k. To this end, we need to define quantities λ_k:

739. Exercise. Write down an expression for numbers $\lambda_k > 0$ having the property: if $A \subset [0, n_{k+1})$ has relative density at most ϵ_{k+1}, then if we break $[0, n_{k+1})$ into sub-intervals of length n_k, restricted to at least $\lambda_k \frac{n_{k+1}}{n_k}$ of these sub-intervals, A has relative density at most ϵ_k. •

Inductively define R_k as follows: R_0 consists of those paths p of length n_0 that see all of the scenery in locations $[-n_0^{\frac{1}{3}}, n_0^{\frac{1}{3}}]$. Having defined R_k, R_{k+1} consists of those paths p of length n_{k+1} having the property that if we break $[0, n_{k+1})$ into sub-intervals $I_j = [jn_k, (j+1)n_k)$ of length n_k and choose any $\lambda_k \frac{n_{k+1}}{n_k}$ of these sub-intervals, we can find r and t such that the patch of scenery p visits during times I_r is disjoint from the patch of scenery p visits during times I_t and such that $p|_{I_r}$ and $p|_{I_t}$ are members of R_k.

740. Exercise. Prove that for k large, most paths of length n_k are reasonable. *Comment: this requires a bit of knowledge about random walks; since we aren't giving any hints, the demand on the reader is high.* •

741. Exercise. Prove that $A_0 \le 2^{-2n_0^{\frac{1}{3}}} = 2^{-2 \cdot 10^6}$. •

Putting $C_k = \frac{A_{k+1}}{A_k^2}$ gives us our required recurrence relation. *Claim:* $C_k < 4n_{k+1}^4 n_k^2$.

742. Exercise. Use the claim to establish that $\sum_{n=0}^{\infty} \frac{\log C_n}{2^{n+1}} < 10^6$. Use Exercises 738 and 741 to show that $A_k \to 0$, completing the proof. •

We prove the claim. Fix a scenery s_1 and a path $p_2 \in R_{k+1}$. For choices r, B and b, where $0 \le r < \frac{n_{k+1}}{n_k}$, B is an interval of length at most n_k such that $B \cap [-rn_k, rn_k] \ne \emptyset$, and b is an even number in $B \cap [-rn_k, rn_k]$, let $E_{r,B,b}$ be the event that randomly chosen scenery on the interval B has the property that if p_1 is a path of length rn_k that terminates at scenery location

b, then p_1 can be extended through times $I_r = [rn_k, (r+1)n_k)$ in such a way that p_1 doesn't venture outside of B throughout times I_r and such that $d_{I_r}\big((s_1, p_1), (s_2, p_2)\big) < \epsilon_k$. Clearly $P(E_{r,B,b}) \le A_k$. Hence, letting $E_B = \bigcup_{r,b} E_{r,B,b}$, one has $P(E_B) \le \frac{n_{k+1}}{n_k} n_k A_k = n_{k+1} A_k$.

Consider next the event E that a randomly chosen scenery s_2 is such that there exists a path p_1 with $d_{[0,n_{k+1})}\big((s_1, p_1), (s_2, p_2)\big) < \epsilon_{k+1}$. For such s_2 and corresponding p_1, we can find at least $\lambda_k \frac{n_{k+1}}{n_k}$ sub-intervals $I_j = [jn_k, (j+1)n_k)$ with $d_{I_j}\big((s_1, p_1), (s_2, p_2)\big) < \epsilon_k$. Since $p_2 \in R_{k+1}$ some two of these, say I_r and I_t, will be such that $p_2|_{I_r}$ and $p_2|_{I_t}$ belong to R_k, and such that the blocks of scenery B_r and B_t visited by p_2 during times I_r and I_t respectively are disjoint. Notice that this implies that E_{B_r} and E_{B_t} are independent.

743. Exercise. Show that $s_2 \in (E_{B_r} \cap E_{B_t})$.　　　　　　　　•

It follows from the previous exercise that $E \subset \bigcup_{B_r, B_t} (E_{B_r} \cap E_{B_t})$. Since there are at most $2n_{k+1}n_k$ choices for each of B_r and B_t, this yields $P(E) \le 4n_{k+1}^4 n_k^2 A_k^2$. Taking the maximum over all s_1 and reasonable P_2 yields $A_{k+1} \le 4n_{k+1}^4 n_k^2 A_k^2$, as desired. □

Appendix: Some open problems

We close the book with a "top ten list". Ten challenging open problems, a solution to any one of which could greatly help to revitalize isomorphism theory. The list isn't supposed to be exhaustive, nor are the problems on it necessarily the most important ones in isomorphism theory; they reflect authorial interest.

744. Definition. A stationary process $(X_i)_{i=-\infty}^{\infty}$ is said to be *loosely Bernoulli* if the fbar distance between the n-future conditioned on the past and the unconditioned n-future goes to zero in probability as $n \to \infty$.

745. Comment. Alternatively, $(X_i)_{i=-\infty}^{\infty}$ is loosely Bernoulli if for a.e. past p and every $\epsilon > 0$ there is an n such that the n-future conditioned on p is within ϵ of the unconditioned n-future in fbar.

746. Problem 1. Let (X, T) denote Chacon's transformation (see Definition 687). Is $(X \times X, T \times T)$ loosely Bernoulli?

747. Comment. Note that Chacon's transformation is rank one, and hence has zero entropy, as does its product. It may be useful to know, therefore (and the reader may take it as an exercise to show), that a zero entropy transformation is loosely Bernoulli if and only if for any two names outside a set of measure 0, the fbar distance of the initial n names approaches 0 as $n \to \infty$. Alternatively, a zero entropy transformation is loosely Bernoulli if and only if it is Kakutani equivalent to an irrational rotation of the circle. (Two systems (X, T) and (Y, S) are Kakutani equivalent if there are sets of equal measure, $A \subset X$ and $B \subset Y$, such that T_A is isomorphic to S_B, where T_A and S_B denote induced transformations; see Definition 447.)

748. Definition. A measure-preserving system (X, T) has *Lebesgue spectrum* if there is a set $F \subset L^2(X)$ such that $\{1\} \cup \{T^n f : f \in F, n \in \mathbf{Z}\}$ forms an orthonormal basis for $L^2(X)$. If F consists of a single function f, we say that (X, T) has *simple Lebesgue spectrum.*

749. Problem 2. Does there exist a system (X, T) having simple Lebesgue spectrum?

750. Comment. Lebesgue spectrum is perhaps the best known variety of mixing we didn't consider in Chapter 7 (there are many, some obscure). Indeed, it lies properly between the K property and strong mixing. That it is unknown whether any systems have simple Lebesgue spectrum seems to us remarkable.

751. Problem 3. (*Furstenberg.*) Let $X = [0, 1)$ and define $T_m : X \to X$ by $T_m x = mx$ (mod 1). If μ is a non-atomic measure on X that is invariant for T_2 and T_3, is μ necessarily Lebesgue measure?

752. Comment. D. Rudolph (1991) has established the foregoing in the case where at least one of (X, μ, T_2), (X, μ, T_3) has positive entropy; we mention also that R. Lyons (1988) proved an important preliminary to this result (for K transformations).

753. Problem 4. Is it the case that if T and S are two K transformations with the same entropy, there is always a way to join T with S to form another K transformation having that same entropy?

754. Problem 5. If T has positive but finite entropy, is T necessarily isomorphic to a process on a finite alphabet Λ such that for every $a \in \Lambda$, $P\bigl(X(0) = a \big| \bigvee_{i=-n}^{-1} X_i\bigr)$ converges uniformly over pasts? If not, can T be extended to such a process?

755. Comment. The zero entropy case of the foregoing problem is known; it is proved by Kalikow *et al.* (1992).

756. Problem 6. Let (X, T) be a strongly mixing system. Is it necessarily the case that for every measurable $A \subset X$, one has $\lim_{n,m,n-m\to\infty} \mu\bigl(A \cap T^{-n}A \cap T^{-m}A\bigr) = \mu(A)^3$? Is it necessarily the case that for every measurable $A \subset X$, one has $\lim_{n\to\infty} \mu\bigl(A \cap T^{-n}A \cap T^{-2n}A\bigr) = \mu(A)^3$?

757. Comment. This problem is generally phrased "does twofold mixing imply threefold mixing?" For rank-one transformations, the answer is *yes*; see Kalikow (1984). Also for mixing transformations with singular spectrum;[110] see Host (1991). We remark that it is unknown whether all rank-one transformations have singular spectrum.

758. Problem 7. Can every positive entropy transformation be written as a join of two Bernoulli transformations?

759. Comment. It is known that every positive entropy transformation can be written as a join of three Bernoulli transformations (Smorodinsky and Thouvenet 1979), or as a factor of a join of two (this is an exercise).

---

[110] This refers to the spectral type of the unitary operator associated with the transformation, a topic that lies outside the scope of this book.

760. Problem 8. If T is not isomorphic to S, can $T \times T$ be isomorphic to $S \times S$?

761. Problem 9. (*Thouvenot; weak Pinsker conjecture.*) Can every ergodic transformation be written as a product of a Bernoulli transformation and an ergodic transformation of arbitrarily small entropy? If not, can every transformation be written as any non-trivial product?

762. Definition. Let $(X, \mathcal{A}, \mu, T)$ be an ergodic system and let $\mathcal{B} \subset \mathcal{A}$ be a factor. Write $\mu = \int \mu_x \, d\mu(x)$ as in Theorem 247. Now put a measure $\mu \times_\mathcal{B} \mu$ on the product space $X \times X$ by the rule $\mu \times_\mathcal{B} \mu = \int (\mu_x \times \mu_x) \, d\mu(x)$. The system $(X, \times X, \mu \times_\mathcal{B} \mu, T \times T)$ is the *relatively independent joining* of (X, T) with itself over the factor $\mathcal{B}$.

763. Problem 10. Can a relatively independent self-joining of a Bernoulli transformation over one of its factors be ergodic but not K?

References

T. Adams, 1998. Smorodinsky's conjecture on rank one mixing, *Proc. Amer. Math. Soc.* **126**, 739–744.

G.D. Birkhoff, 1931. Proof of the ergodic theorem, *Proc. Nat. Acad. Sci.* **17**, 656–660.

L. Breiman, 1957. The individual theorem of information theory, *Ann. Math. Stat.* **28**, 809–811; errata 1960, **31** 809–810.

R.M. Dudley, 2002. *Real Analysis and Probability*, Cambridge University Press.

G. Folland, 1984. *Real Analysis. Modern Techniques and their Applications. Pure and Applied Mathematics.* John Wiley, New York.

H. Furstenberg, 1967. Disjointness in ergodic theory, minimal sets and a problem in Diophantine approximation, *Math. Systems Theory* **1**, 1–49.

H. Furstenberg, 1981. *Recurrence in Ergodic Theory and Combinatorial Number Theory.* M. B. Porter Lectures. Princeton University Press.

H. Furstenberg and Y. Katznelson, 1985. An ergodic Szemerédi theorem for IP-systems and combinatorial theory, *J. Analyse Math.* **45**, 117–168.

H. Furstenberg and B. Weiss, 1977. The finite multipliers of infinite ergodic transformations. *The Structure of Attractors in Dynamical Systems (Proc. Conf., North Dakota State Univ., Fargo, N.D., 1977), Lecture Notes Math.*, **668**, 127–132.

R.M. Gray, 1988. *Probability, Random Processes, and Ergodic Properties*, Springer-Verlag, New York.

R.M. Gray, 1991. *Entropy and Information Theory*, Springer-Verlag, New York.

N. Hindman, 1974. Finite sums from sequences within cells of a partition of N, *J. Combinatorial Theory Ser. A*, **17**, 1–11.

C. Hoffman and D. Rudolph, 2002. Uniform endomorphisms which are isomorphic to a Bernoulli shift, *Ann. Math. (2)* **156**, 79–101.

B. Host, 1991. Mixing of all orders and pairwise independent joinings of systems with singular spectrum, *Israel J. Math.* **76**, 289–298.

S. Kalikow, 1982. T, T^{-1} transformation is not loosely Bernoulli, *Ann. Math. (2)* **115**, 393–409.

S. Kalikow, 1984. Twofold mixing implies threefold mixing for rank one transformations, *Ergodic Theory Dynam. Systems* **4**, 237–259.

S. Kalikow, Y. Katznelson and B. Weiss, 1992. Finitarily deterministic generators for zero entropy systems, *Israel J. Math.* **79**, 33–45.

J. Kieffer, 1984. A simple development of the Thouvenot relative isomorphism theory, *Ann. Prob.* **12**, 204–211.

A.N. Kolmogorov, 1958. A new metric invariant of transitive dynamical systems and Lebesgue space automorphisms, *Dokl. Acad. Sci. USSR* **119**, 861–864.

U. Krengel, 1985. *Ergodic Theorems.* With a supplement by Antoine Brunel. *de Gruyter Studies in Mathematics* **6**, Walter de Gruyter, Berlin.

W. Krieger, 1970. On entropy and generators of measure preserving transformations, *Trans. Amer. Math. Soc.* **149**, 453–464.

R. Lyons, 1988. On measures simultaneously 2- and 3-invariant, *Israel J. Math.* **61**, 219–224.

R. McCutcheon, 1999. *Elemental Methods in Ergodic Ramsey Theory,* Lecture Notes in Mathematics **1722**. Springer-Verlag, Berlin.

B. McMillan, 1953. The basic theorems of information theory, *Ann. Math. Stat.* **24**, 196–219.

D. Ornstein, 1970*a*. Bernoulli shifts with the same entropy are isomorphic, *Adv. Math.* **4**, 337–352.

D. Ornstein, 1970*b*. Bernoulli shifts with infinite entropy are isomorphic, *Adv. Math.* **5**, 339–348.

D. Ornstein, 1973. An example of a Kolmogorov automorphism that is not a Bernoulli shift, *Adv. Math.* **10**, 49–62.

D. Ornstein, 1974. *Ergodic Theory, Randomness, and Dynamical Systems,* Yale University Press, New Haven, CT.

D. Ornstein and B. Weiss, 1974. Finitely determined implies very weak Bernoulli, *Israel J. Math.* **17**, 94–104.

S.T. Rachev and L Rüschendorf, 1998. *Mass Transportation Problems. Volume I: Theory. Volume II: Applications*, Springer, New York.

V. A. Rohlin, 1952. On the fundamental ideas of measure theory, *Amer. Math. Soc. Transl.* **1952**, no. 71.

V. A. Rohlin, 1965. Generators in ergodic theory. II. *Vestnik Leningrad. Univ.* **20**, 68–72.

D. Rudolph, 1990. *Fundamentals of Measurable Dynamics. Ergodic Theory on Lebesgue Spaces.* Oxford University Press.

D. Rudolph, 1991. ×2 and ×3 invariant measures and entropy, *Ergodic Theory and Dynamical Systems* **10**, 395–406.

C. Shannon, 1948. A mathematical theory of communication, *Bell Syst. Tech. J.* **27**, 379–423, 623–656.

P. Shields, 1979. Almost block independence, *Prob. Th. Related Fields* **49**, 119–123.

P. Shields, 1996. *The Ergodic Theory of Discrete Sample Paths*, Graduate Studies in Mathematics, **13**. American Mathematical Society.

Y. Sinai, 1964. On weak isomorphism of transformations with invariant measure, *Math. Sb. (N.S.)*, **63 (105)**, 23–42.

M. Smorodinsky and J. Thouvenot, 1979. Bernoulli factors that span a transformation, *Israel J. Math.* **32**, 39–43.

J. Thouvenot, 2002. Entropy, isomorphism and equivalence in ergodic theory, *Handbook of Dynamical Systems*, Volume 1A. North-Holland, Amsterdam, pp. 205–238.

P. Walters, 2000. *An Introduction to Ergodic Theory*, Springer-Verlag, New York.

Index

Please note that entries are numbered by paragraph, e.g. the "extended monkey method" can be found in paragraph 266 on page number 54.

Printed in the United States
by Baker & Taylor Publisher Services

Printed in the United States
by Baker & Taylor Publisher Services